T0211828

Lecture Notes
in Business Information Processing   430

More information about this series at http://www.springer.com/series/7911

Robert Andrei Buchmann · Andrea Polini ·
Björn Johansson · Dimitris Karagiannis (Eds.)

Perspectives in Business Informatics Research

20th International Conference
on Business Informatics Research, BIR 2021
Vienna, Austria, September 22–24, 2021
Proceedings

Editors
Robert Andrei Buchmann 🆔
Babeş-Bolyai University
Cluj Napoca, Romania

Björn Johansson 🆔
Linköping University
Linköping, Sweden

Andrea Polini 🆔
University of Camerino
Camerino, Italy

Dimitris Karagiannis 🆔
University of Vienna
Vienna, Austria

ISSN 1865-1348   ISSN 1865-1356 (electronic)
Lecture Notes in Business Information Processing
ISBN 978-3-030-87204-5   ISBN 978-3-030-87205-2 (eBook)
https://doi.org/10.1007/978-3-030-87205-2

This Springer imprint is published by the registered company Springer Nature Switzerland AG
The registered company address is: Gewerbestrasse 11, 6330 Cham, Switzerland

Preface

The 20th anniversary of the International Conference on Perspectives in Business Informatics Research (BIR 2021) was organized by the University of Vienna and held during September 22–24, 2021.

This is an important milestone in the history of European Business Informatics research, highlighted with a retrospective keynote presentation. It confirms BIR as one of the flagship conferences for this interdisciplinary field, which is at the intersection of Business Administration and Computer Science. BIR was born from a workshop series launched in the late 90s and since then it has provided a forum for scientific inquiry—primarily interested in the convergence of innovative business ecosystems and rapidly shifting technological paradigms. Business-IT alignment is therefore a core concern of the BIR community, together with its supporting methods, both conceptual and computational.

This year, the main conference received 49 submissions out of which 16 were selected after a review process involving at least 3 reviewers for each paper. This translates to a 33% acceptance rate for the main conference proceedings. The papers are grouped into the following topics: Technology Adoption and Acceptance during COVID-19 Times, Conceptual Modeling for Enterprise Systems, Enterprise Modeling Methods and Frameworks, Compliance and Normative Challenges, and Empirical Investigations on Digital Innovation and Transformation Prerequisites.

These topical headings indicate a balance between design-oriented research and empirical investigation, with the second category being boosted by challenges emerging from the pandemic conditions, especially in educational (e-learning) contexts. In the future, we hope to have less of the real challenge and more preventive or reactive treatments which must, of course, be informed by the observational research reported during these times. By balancing the empirical and design-oriented research, BIR 2021 advocates a unified view on the two types of research, as opposed to the more prominent dichotomous perception of the two approaches—empirical observation should inform innovative design and innovative design should be tweaked according to empirical experimentation. Conference and workshop papers may limit their scope to one or the other of these phases, but it is an explicit goal of conferencing to foster discussions that reveal bridges between these papers and will hopefully inspire the methodologically holistic research that we need in Business Informatics.

Traditionally, the conference has included a number of satellite events which help promote its status as a scientific networking hub that stimulates critical discussion and knowledge exchange on topics related to Business Informatics. This year, three workshops were open to submissions: the 6th edition of Managed Complexity (ManComp), the 12th edition of Information Logistics and Digital Transformation (ILOG) and the 1st edition of Blockchain for Trusted Data Sharing (B4TDS). In addition, the traditional doctoral consortium section attracted doctoral papers subjected to mentoring and recommendations that will prove insightful for junior researchers. The BIR workshops and the doctoral consortium publish a joint proceedings volume in the CEUR-WS series.

We express our gratitude to the BIR Steering Committee members who agreed to have BIR 2021 managed by our team. In particular, we thank Prof. Marite Kirikova and Prof. Kurt Sandkuhl for their decision-making assistance and for acting as this year's workshop chairs.

We thank all the authors who submitted their work and the Program Committee members who contributed timely reviews to the paper selection process. Many strong papers were submitted and the selection, aiming to preserve the traditional acceptance threshold of BIR, was particularly difficult this year.

We also thank the Open Models Laboratory (OMiLAB, www.omilab.org) community of practice for contributing either submissions or reviews to this anniversary edition of BIR.

The Springer team lead by Ralf Gerstner and Christine Reiss again provided prompt support regarding the proceedings production. We are thankful to them for the continuous partnership with BIR, which is a major factor for the conference longevity and the prestige of its proceedings.

Last but not least, the University of Vienna team led by Victoria Döller deserves congratulations for managing the successful event, which this year hosts presentations for papers accepted during both the 2020 and 2021 editions as well as the satellite events that had to be postponed last year.

August 2021   Björn Johansson
                                                                     Dimitris Karagiannis
                                                              Robert Andrei Buchmann
                                                                         Andrea Polini

Organization

BIR 2021 was organized and hosted at the University of Vienna, Austria, during September 22–24, 2021.

General Chairs

Björn Johansson	Linköping University, Sweden
Dimitris Karagiannis	University of Vienna, Austria

Program and Publication Co-chairs

Robert Andrei Buchmann	Babeş-Bolyai University of Cluj Napoca, Romania
Andrea Polini	University of Camerino, Italy

Local Organizing Chair

Victoria Döller	University of Vienna, Austria

Steering Committee

Marite Kirikova (Chair)	Riga Technical University, Latvia
Björn Johansson (Vice-chair)	Linköping University, Sweden
Eduard Babkin	State University - Higher School of Economics, Nizhny Novgorod, Russia
Rimantas Butleris	Kaunas Technical University, Lithuania
Sven Carlsson	Lund University, Sweden
Peter Forbrig	University of Rostock, Germany
Andrzej Kobyliñski	Warsaw School of Economics, Poland
Lina Nemuraite	Kaunas Technical University, Lithuania
Jyrki Nummenmaa	University of Tampere, Finland
Raimundas Matulevicius	University of Tartu, Estonia
Vaclav Repa	Economic University Prague, Czech Republic
Kurt Sandkuhl	University of Rostock, Germany
Benkt Wangler	University of Skövde, Sweden
Stanislaw Wrycza	University of Gdansk, Poland

Program Committee

Jan Aidemark	Linnaeus University, Sweden
Asif Akram	Chalmers University of Technology, Sweden
Gundars Alksnis	Riga Technical University, Latvia
Bo Andersson	Lund University, Sweden
Said Assar	Institut Mines-Télécom Business School, France
Eduard Babkin	National Research University Higher School of Economics, Russia
Per Backlund	University of Skövde, Sweden
Amelia Bădică	University of Craiova, Romania
Peter Bellström	Karlstad University, Sweden
Cătălin Boja	Bucharest University of Economic Studies, Romania
Dominik Bork	Technical University of Vienna, Austria
Tomáš Bruckner	University of Economics and Business in Prague, Czech Republic
Robert Buchmann	Babeş-Bolyai University, Romania
Witold Chmielarz	University of Warsaw, Poland
Michael Le Duc	Mälardalen University, Sweden
Blerim Emruli	Lund University, Sweden
Umberto Fiaccadori	Lund University, Sweden
Hans-Georg Fill	University of Fribourg, Switzerland
Peter Forbrig	University of Rostock, Germany
Chiara Di Francescomarino	Bruno Kessler Foundation, Italy
Ahmad Ghazawneh	Halmstad University, Sweden
Ana-Maria Ghiran	Babeş-Bolyai University, Romania
Janis Grabis	Riga Technical University, Latvia
Knut Hinkelmann	University of Applied Sciences Northwestern Switzerland, Switzerland
Nicklas Holmberg	Lund University, Sweden
Adrian Iftene	Alexandru Ioan Cuza University, Romania
Emilio Insfran	Universitat Politécnica de València, Spain
Amin Jalali	Stockholm University, Sweden
Florian Johannsen	University of Applied Sciences Schmalkalden, Germany
Björn Johansson	Linköping University, Sweden
Miranda Kajtazi	Lund University, Sweden
Dimitris Karagiannis	University of Vienna, Austria
Christina Keller	Jönköping University, Sweden
Sybren De Kinderen	University of Duisburg-Essen, Germany
Marite Kirikova	Riga Technical University, Latvia
Markus Lahtinen	Lund University, Sweden
Michael Lang	National University of Ireland, Galway, Ireland
Birger Lantow	University of Rostock, Germany
Moonkun Lee	Chonbuk National University, South Korea
Massimiliano de Leoni	University of Padua, Italy

Ginta Majore   Vidzeme University of Applied Sciences, Latvia
Raimundas Matulevicius   University of Tartu, Estonia
Andrea Morichetta   University of Camerino, Italy
Jacob Nørbjerg   Copenhagen Business School, Denmark
Jyrki Nummenmaa   University of Tampere, Finland
Malgorzata Pankowska   University of Economics in Katowice, Poland
Victoria Paulsson   Dublin City University, Sweden
Jens Myrup Pedersen   Aalborg University, Denmark
Dana Petcu   West University of Timişoara, Romania
Tomas Pitner   Masaryk University, Czech Republic
Paul Pocatilu   Bucharest University of Economic Studies, Romania
Andrea Polini   University of Camerino, Italy
Pierluigi Plebani   Polytechnic University of Milan, Italy
Dorina Rajanen   University of Oulu, Finland
Barbara Re   University of Camerino, Italy
Vaclav Repa   University of Economics in Prague, Czech Republic
Ben Roelens   Open University of Netherlands, The Netherlands
Christian Săcărea   Babeş-Bolyai University, Romania
Kurt Sandkuhl   University of Rostock, Germany
Rainer Schmidt   Munich University of Applied Sciences, Germany
Gheorghe Cosmin Silaghi   Babeş-Bolyai University, Romania
Janis Stirna   Stockholm University, Sweden
Frantisek Sudzina   Aalborg University, Denmark
Ann Svensson   University West, Sweden
Torben Tambo   Aarhus University, Denmark
Filip Vencovsky   University of Economics in Prague, Czech Republic
Gianluigi Viscusi   Imperial College Business School, UK
Anna Wingkvist   Linnaeus University, Sweden
Wiesław Wolny   University of Economics in Katowice, Poland
Stanislaw Wrycza   University of Gdansk, Poland
Jelena Zdravkovic   University of Stockholm, Sweden
Alfred Zimmermann   Reutlingen University, Germany

Additional Reviewers

Natalia Aseeva   State University - Higher School of Economics, Nizhny
                                   Novgorod, Russia
Syed Juned Ali   Technical University of Vienna, Austria
Felix Härer   University of Fribourg, Switzerland
Shahrzad Khayatbashi   Linköping University, Sweden
Alessandro Marcelletti   University of Camerino, Italy
Fabian Muff   University of Fribourg, Switzerland
Achim Reiz   University of Rostock, Germany
Jack Daniel Rittelmeyer   University of Rostock, Germany
Gezim Sejdiu   Deutsche Post DHL Group, Germany

The 20-years Odyssey of the Perspectives in Business Informatics Research (BIR) Conference (Keynote Talk)

Sven Carlsson[1], Björn Johansson[2] and Benkt Wangler[3]

[1] Lund University, Sweden
sven.carlsson@ics.lu.se
[2] Linköping University, Sweden
bjorn.se.johansson@liu.se
[3] Stockholm University, Sweden
benktw@gmail.com

The BIR series of conferences traces its origin back to a collaboration that took place between researchers at universities in Berlin, Rostock (Germany) and Växjö (Sweden) during the late nineties. The group in Växjö was led by professor Bengt G. Lundberg, the initiator of a workshop series on "perspectives in business informatics research". The last workshop in this series was organized in Rostock, 2000, and became actually the starting point of the conference series "Perspectives in Business Informatics Research (BIR). The professors in Berlin, Rostock and Växjö wanted to create a forum where researchers in business informatics, junior as well as senior, could meet and discuss with each other. The primary focus was PhD students and junior researchers. The term Business Informatics was chosen as a translation of the German "wirtschaftsinformatik".

After professor Lundberg passed away 2001, the initiative on the Swedish side was passed on to a research network named Knowledge in Organizations (KiO) also led by professor Lundberg together with two of the authors of this piece of text, Benkt Wangler and Sven Carlsson. The latter two together with their colleagues in Berlin and Rostock, and supported by PhD students, decided to continue with BIR conferences, and a second conference was then organized in Berlin 2003. From that time on the conference has been organized each year with the exception of 2020, when the conference was postponed for one year due to the Covid-19 pandemic.

The original idea was to have a series of seminars/workshops/conferences mainly for PhD students to make it possible for them to write papers, get them evaluated by senior researchers and to present the PhD students' work. The conference should be for two full days, which was later complemented by an initial day of half or full day specialized workshops. It was also in part based on the idea of supporting PhD students and junior researchers at colleges, new universities, and other universities without strong BIR-research traditions. For example, Växjö university did not have a PhD education in business informatics (the students were enrolled at Stockholm University). However, gradually from one year to the next the ambition with the conference, the number of papers submitted and the quality of accepted papers increased. The quality of papers presented in the conference proceeding has also increased by gradually having a lower acceptance rate - the recent years this has been 30-35%. Nevertheless, the conference has always had a generous attitude towards letting junior researchers present their ideas, if so in some special forum session not included in the official proceedings. The first focus has been extended to also include more senior researchers, but the original focus has been important and supported over the years.

Another idea was related to conference themes and research. It was said that each conference could have a specific theme, but that at the same time each conference should be eclectic regarding what to address (not bounded by the conference theme), what research approaches and methods to use, what particular theories to employ, what the philosophical underpinnings are, etc. It was decided that the conference should welcome contributions addressing immediate areas of concern (e.g., as stated in a conference's theme), situational areas of concern (e.g., business-to-business commerce), and classical areas of concern (e.g., the systems development process). Looking at the accepted papers we conclude that the BIR conferences have fulfilled the "conference themes and research" idea well.

A third idea was related to "after working hours" activities. The intention was that the conferences should arrange informal after working hours activities, not only a formal welcome reception and a conference dinner. The intention was to make it possible for more informal networking. This has been fulfilled in different forms, for example, visits to breweries with tasting, boat trips, visits to museums and guided city tours.

Gradually it also became evident that albeit the conferences welcome papers from all over the world, BIR had its domicile in the Baltic Sea region, a fact that was also explicitly stated in the conference policy. However, it was also stated that this should not hinder that it is sometimes organized elsewhere, as long as it frequently returns to the Baltic Sea region. The conference is always organized by some university local to the place where the conference is held. Looking backwards, the conference has stayed in the Baltic Sea region, but on a few occasions it has also been organized outside this area. Table 1 lists the conferences and where they were held.

At the conference in **Kaunas 2006** a formal steering committee for the conference series was set up. Benkt Wangler was elected chairman, a position that he held until 2009. Stanislaw Wrycza was then chairman 2009–2010, Kurt Sandkuhl 2010–2016 and since 2016 Marite Kirikova has been the chair of the steering committee. When Marite Kirikova was elected as chair it was also decided to have a vice chair, a position that fell upon Björn Johansson 2016. The same two persons still hold these positions.

In **Rostock 2010** the conference was complemented with a doctoral consortium to give PhD students the opportunity to present papers and PhD project ideas, and to get feedback from senior researchers on their overall PhD project. The organizers of that conference also managed to get acceptance for the BIR proceedings to be published with Springer. Hence, starting from the conference in **Rostock 2010**, the conference proceedings are published in the Springer series Lecture Notes in Business Information Processing (LNBIP). Before Springer proceedings papers were published in local proceedings according to each organizing university's policy. Occasionally complementary proceedings have also been published as CEUR Proceedings (CEUR-WS.org).

Table 1. BIR conferences 2003–2021

Year	Conference place	Organizing university	Conference chair/s	Proceedings editor/s	Number of accepted papers
2003	Berlin	Humboldt University of Berlin and University of Potsdam, Germany	Bernd Viehweger	Bernd Viehweger	18 full papers
2004	Rostock	University of Rostock, Germany	Hans Röck	Hans Röck	17 full papers
2005	Skövde	University of Skövde, Sweden	Benkt Wangler	Per Backlund, Sven Carlsson, Eva Söderström	13 full papers
2006	Kaunas	Kaunas University of Technology, Lithuania	Rimantas Butleris	Lina Nemuraité, Benkt Wangler, Rita Butkiené	18 full papers plus 6 poster papers
2007	Tampere	University of Tampere, Finland	Jyrki Nummenmaa	Jyrki Nummenmaa; Eva Söderström	14 full papers
2008	Sopot	University of Gdansk, Poland	Stanislaw Wrycza	Stanislaw Wrycza	35 full papers
2009	Kristianstad	Kristianstad University, Sweden	Harald Kjellin	Jan Aidemark, Sven Carlsson, Björn Cronquist	20 full papers plus 6 short papers
2010	Rostock	University of Rostock, Germany	Peter Forbrig, Horst Gunther	Peter Forbrig, Horst Gunther	14 full papers plus 4 short papers published in the Springer LNBIP series.
2011	Riga	Riga Technical University, Latvia	Janis Grabis, Marite Kirikova	Janis Grabis, Marite Kirikova	25 full papers published in LNBIP

(continued)

Table 1. (*continued*)

Year	Conference place	Organizing university	Conference chair/s	Proceedings editor/s	Number of accepted papers
2012	Nizhny Novgorod	NRU HSE in Nizhny Novgorod, Russia	Oleg Kozyrev	Natalia Aseeva, Eduard Babkin, Oleg Kozyrev	15 full papers published in LNBIP
2013	Warsaw	Warsaw School of Economics, Poland	Andrzej Kobyliński, Andrzej Sobczak	Andrzej Kobyliński, Andrzej Sobczak	19 full papers and 5 short papers published in LNBIP
2014	Lund	Lund University, Sweden	Björn Johansson	Björn Johansson, Bo Andersson, Nicklas Holmberg	27 full papers published in LNBIP
2015	Tartu	University of Tartu, Estonia	Raimundas Matulevičius, Marlon Dumas	Raimundas Matulevičius, Marlon Dumas	16 full papers and four short papers published in LNBIP
2016	Prague	University of Economics, Czech Republic	Václav Řepa, Tomáš Bruckner	Václav Řepa, Tomáš Bruckner	22 full papers and two short papers published in LNBIP
2017	Copenhagen	Aalborg University, Denmark	Björn Johansson, Charles Møller	Björn Johansson, Charles Møller, Atanu Chaudhuri, Frantisek Sudzina	17 full papers and three short papers published in LNBIP
2018	Stockholm	Stockholm University, Sweden	Selmin Nurcan, Janis Stirna	Jelena Zdravkovic, Jānis Grabis, Selmin Nurcan, Janis Stirna	17 full papers published in LNBIP
2019	Katowice	University of Economics in Katowice, Poland	Małgorzata Pańkowska, Kurt Sandkuhl	Małgorzata Pańkowska, Kurt Sandkuhl	17 full papers published in LNBIP

(*continued*)

Table 1. (*continued*)

Year	Conference place	Organizing university	Conference chair/s	Proceedings editor/s	Number of accepted papers
2020	No physical event (presentations merged with the 2021 event)	University of Vienna, Austria	Björn Johansson, Dimitris Karagiannis	Robert Andrei Buchmann, Andrea Polini, Björn Johansson, Dimitris Karagiannis	14 full papers published in LNBIP
2021	Vienna	University of Vienna, Austria	Björn Johansson, Dimitris Karagiannis	Robert Andrei Buchmann, Andrea Polini, Björn Johansson, Dimitris Karagiannis	16 full papers published in LNBIP

From the start until 2009 presented papers were published in local proceedings.
From the conference in 2010 accepted papers have been published by Springer in Lecture Notes in Business Information Processing (LNBIP) in the series "Perspectives in Business Informatics Research".
The LNBIP BIR proceedings can be found at
https://link.springer.com/conference/bir

At the conference in **Riga 2011**, the organizers introduced a pre-conference day that included workshops and the doctoral consortium. The team of this conference also did a great job in attracting papers and participants. So far, the **Riga 2011** conference has had the highest number of both papers and participants. At the conference there were four workshops organized during a pre-conference day. Since that the BIR conference has included workshops on topics such as: Information logistics and digital transformation, Alignment of business process and security modeling, Intelligent educational systems, User oriented information integration, Ontologies and information systems, Business processes in the cloud, Managed complexity, Security and compliance in business processes, Resilient enterprise architecture and Security analytics.

In **Lund 2014** there was for the first time a keynote by an industry representative. This was a highly appreciated initiative. The keynote organization was also involved in two special workshops that had the character of being more hands-on. The workshops had students from the informatics program (both bachelor and master) as well as junior and senior researchers attending the conference as participants. This interaction between students and researchers was highly appreciated. To the best of our knowledge the Lund 2014 conference is so far the only conference that have had students listening to paper presentations at the main conference.

In **Copenhagen 2017** the pre-conference day included a pre-BIR forum day, in which borderline papers that did not fit into the organized workshops were presented and discussed. For each paper discussants were appointed who read the papers beforehand and the main focus was on discussing papers. This was seen, from the participants, as a very good way of getting more substantial feedback and a good discussion around the individual paper. This is something that future BIR conferences could benefit from doing again.

In 2020 the organizer decided, due to the pandemic situation, to postpone the "happening". This decision was made from the fact that the BIR conference – with its satellite events – has always been a scientific networking hub, encouraging elaborated critical discussion and providing knowledge exchange opportunities. The organizers did not see that this could happen if having the conference as a virtual event. Therefore, the organization team decided to postpone BIR 2020 event but still create the 2020 proceedings and have the papers presented as part of an extended event to take place during BIR 2021.

Over the years, the conference has gained in reputation and interest among researchers in Business Informatics. Gradually from one year to the next the ambition with the conference, and the number of papers submitted and the quality of accepted papers has increased. At the same time the initial focus of having the conference support PhD students and junior researchers in their academic journey in a friendly and stimulating environment has been kept. The ambition is that future conferences will continue to do so.

Contents

Compliance and Normative Challenges

**Empirical Investigations on Digital Innovation and Transformation
Prerequisites**

Technology Adoption and Acceptance During COVID-19 Times

Impact of the Pandemic on the Barriers to the Digital Transformation in Higher Education - Comparing Pre- and Intra-Covid-19 Perceptions of Management Students

Sven Packmohr[1]([⊠]) [iD] and Henning Brink[2] [iD]

[1] Malmö University, 205 06, Malmö, Sweden
sven.packmohr@mau.se
[2] Osnabrück University, 49069 Osnabrück, Germany
henning.brink@uni-osnabrueck.de

Abstract. The rise of digital technologies is a macro trend, forcing organizations to transform digitally. This so-called digital transformation (DT) is affecting the field of higher education, too. Higher education institutions (HEI) digitalize internal processes and offer digitally-enabled education services. Different types of barriers are challenging a successful DT and need to be mastered. Our study follows a longitudinal research design by surveying different student cohorts in the same courses. Before the pandemic, we identified the barriers to DT and transferred them into a research model. Pre-pandemic, we surveyed the influence of barriers perceived by management students on the DT process of their HEI. Taking the pandemic as a solid external driver on DT, we examined students' intra-pandemic perception in the same courses as the pre-pandemic analysis. With pre-pandemic data, the projection explains over 50% of the adjustment problems of the DT process. Based on intra-pandemic data, the explanation decreases to 45%. Hypothetically, we expected a better explanation degree as an impact of the pandemic. Interestingly, results indicate that intra-pandemic perceptions got more complex and, therefore, less significant.

Keywords: Digital transformation · Barriers · Student perception · Covid-19 · Higher education

1 Introduction

The concept of digital transformation (DT) aims at using new digital technologies to enable significant improvements in organizations at different levels, such as processes, business models, and connections with different stakeholders [1]. All areas of society are impacted by DT, such as teaching and learning in higher education (HE) [2]. DT in HE brings a shift to digital teaching and learning methods as well as digital administrative processes. Digital technologies allow for constructivist learning and learner assessment. They are transforming access to learning materials, dialogue with, and collaboration

© Springer Nature Switzerland AG 2021
R. A. Buchmann et al. (Eds.): BIR 2021, LNBIP 430, pp. 3–18, 2021.
https://doi.org/10.1007/978-3-030-87205-2_1

between involved groups. The potential is promising, but higher education institutions (HEIs) face obstacles in making sense of these technologies and integrating them into an overall package. Although digital content has been used in virtual learning environments for years, its use was still spotty before the Covid-19 pandemic. Obstacles exist to the full integration and alignment of digital content into degree programs [3]. Legal and organizational procedures, as well as personal biases, slow down DT. We call these obstacles barriers to DT and define them as "those things that hinder, slow down, or stop the DT process" [4].

Due to the Covid-19 pandemic, this transformation process has received a tremendous push forward [5] and forced HEIs to evolve [6]. Existing courses had to be adapted to online-only. Departments had to find new ways to maintain their services under contact restrictions in a short period of time. This sudden change was also accompanied by new ways of working, such as virtual communication [6]. However, the forced DT is not free of barriers [8].

Even before the pandemic, we developed a barrier model using mixed methods [9]. This model contains an applicable scale to measure existing barriers among different stakeholders and follows a socio-technical perspective [10]. In this study, we use the model along with the scale to examine barriers perceived to exist by users in HE before and during the pandemic. Identifying and evaluating these barriers enables improvement of the ongoing DT process. Thus, instructors and administrators at universities can derive suggestions for improvement and work around barriers.

As a sample for our study, we chose to survey management students. They encounter a DT environment as users. In general, students represent the user group in HEIs. Especially management students are particularly critical and aware of their future employability. Even if management students are at another stage of their careers compared to working managers, they share a similar attitude and are trained in certain models such as the Deming circle. Also, these students have grown up as digital natives. Thus, they encounter DT in their daily lives and their education. Also, they will face DT's impact in their professional lives later on [11]. Therefore, DT is thought to be a central topic in the curriculum of management studies [12]. Therefore, our research question is: What differences in the barriers to DT of their HEIs do management students perceive between a pre- and an intra-pandemic educational setting?

We use an existing quantitative instrument to answer this question and survey students at a pre- and intra-pandemic time point when courses were switched from a blended to a complete online design. With this externally forced change, also the DT is enforced. Our hypothesis is that the pandemic will have an effect on the HEIs to overcome barriers. Therefore, we examine the influence of individual, organizational, technical, and environmental barriers on the perception of the DT process by using two different types of digitally organized teaching.

2 Theoretical Development

Increasing technology use and the availability of information assets based on ubiquitous connectivity are currently shaping HE [13]. Digital resources are supporting the conduction of lectures as they enrich learning content. Furthermore, they help with the implementation of strategies and evaluations within learning and teaching [14].

According to the overall concept of DT, it is more than the mere use of technologies for teaching in HEIs. Also, DT implies that data, such as student results, become less private and more traceable. Potentially, lecturers and administrators can easily share data on students. An essential aim for universities under increasing competition is the introduction of digitalized processes, as these are supposed to be more efficient. The usage of digital technologies varies. It ranges from digitally-enabled 1:1 communication between students and teachers to 1:many massive open online courses [15]. In general, DT enhances the possibilities for learning. It adds channels and new forms of content. Thus, blended learning environments have established themselves as further developments of traditional lectures. Whereas even more advanced e-learning offers follow a complete digital design. New approaches promise more positive learning habits as well as attitudes towards learning [16]. However, barriers to the implementation and usage of blended and e-learning designs exist [3, 17]. Initial costs and set-up times for setting up e-learning offers, especially for high-quality content of lecturers, are high. Other problems are organizational interfaces within study programs as well as the access to and the usage of digital technologies [3].

Research on DT in HE mostly addresses certain learning settings. Three groups of HE research exists, which focus on the challenges and gains of DT. The first group is focusing on analyzing student's technology acceptance [18] and DT's effects on students' learning outcomes [19]. The second group contains research on DT-related instructional design and its acceptance [20]. The third group focuses on organizational obstacles within HEIs [21, 22] as these often provoke resistance to change within institutions [23], such as additional workload [24], lack of institutional support [22, 23], and resources such as time and technical equipment. As DT is also impacting the curricula, faculty is often critical to accept these changes [25]. Also, external barriers hinder the DT of HE. Globalization and a more competitive environment are putting pressure on HEIs. However, they are often slow in keeping up with the speed of the change [25].

Recent research on the impact of the pandemic on HE is spreading out in the above-mentioned groups. Marinoni et al. [8] report problems in communication, lack of resources, and problems in pursuing educational and research tasks. Other studies compare data from different time points and find more continuous studying habits among students, leading to higher study efficiency [26].

In conclusion, research on HE is often focusing on specific teaching scenarios. Due to this limited generalizability, we broaden the perspective and take further dimensions of barriers into account to measure the students' view on DT in HE [9].

3 Longitudinal Trend Research Design

In this study, we use a quantitative instrument that was developed according to an exploratory sequential mixed methods approach [27]. Mixed methods are advantageous when complex issues such as policies, interventions, and transformations are researched. They allow for more robust analysis and add details to complex phenomena [28]. For the explorative step, we collected data from 46 interviewees involved in the DT of companies. This data was coded and clustered in different steps to identify the specific dimensions of barriers that might affect the DT process. By using results from the literature, the dimensions were specified into items adopted to a HE setting to survey students and their perceptions. The result is an instrument modeling the causes of DT barriers in HEIs. On a generic level, barriers in HE are not that different from those faced for DT in business settings [29]. We assume that students in HE are getting a taste of how DT will shape future work environments. Especially management education is thought to connect to theoretical models and practical assignments [30]. The transition from management interviews [4] to students' perception is a valid approach as the role of students will evolve over time. Currently, they are using their university. Thus, they are a stakeholder group grown up as digital natives. In the future, they will be digitally involved in their workplaces. Their current perception will influence their attitude toward DT in the long term as future decision-makers. Therefore, we use different dimensions of barriers to DT stemming from business research and apply them in our research on management students.

We use the quantitative instrument to survey two cohorts of management students in the same courses at a time point just before the pandemic and at a time point during the pandemic. The curricula of the programs did not change. Thus, we conduct a comparative longitudinal trend study [31]. The pre-pandemic measurement was taken after the spring term 2019 in the courses Business Process Management and Digital Transformation. Additionally, data from students of previous semesters in the course Project Management were surveyed. All the courses were electives in Business Studies programs and related to the field of information systems in terms of content. Before the pandemic, the courses were instructed with a blended-learning approach of digital components and a supplementary attendance part. Also, students experienced digitalized administrative procedures such as course subscriptions, exam registrations, and communication. One hundred four respondents completed the first round of the questionnaire, with 58.5% male and 41.5% female participants. None indicated a third gender. 80% have already gained initial work experience. The pre-pandemic measurement was taken to answer how students perceive the DT of their HEI. During the pandemic, the hypothesis evolved that it has a positive influence on the perception of DT, as enforced digitally-transformed courses are becoming the new normal. In other areas of HE, positive effects were found [32]. Therefore, we surveyed the next cohort of students in the same three courses at an intra-pandemic time point during the autumn term 2020/2021. One hundred thirteen respondents completed the second round of the questionnaire. The distribution of gender is 66% male and 34% female. None indicated a third gender. Around 66% possess initial work experience.

To compare the two data sets, we examine them statistically by using multiple linear regression analysis. We analyze the effect barriers have on the perceived DT process in both samples.

4   Data Collection Instrument

As described, we conducted a survey at two different time points using the same questionnaire [9]. The questionnaire is displayed in Table 1 and consists of six dimensions measuring the DT Process and its barriers by using a total of 30 items.

Considering that barrier models in literature often formulate the adoption of technology as the target variable [33], we defined the DT process of HEIs as the first and dependent variable in our study. HEIs are becoming digitally-enabled organizations with the need to digitize internal processes on the one hand and develop a broad portfolio of digital services and smart teaching on the other hand [34]. The offering and implementation of smart products and services can be operationalized, observed, and understood as the progression in DT maturity. These different aspects are represented in six items to measure the DT process. We assume that the dependent variable is negatively affected by barriers, the independent variables in our model.

Following the socio-technical perspective, barriers can have an impact on different levels and presumably with different intensities. Based on qualitative interviews, statements about barriers were aggregated into five barrier dimensions. The dimensions allow the individual effect on DT to be examined.

Individual barriers are one of the dimensions. They reflect an individual's difficulties accepting the DT process [4]. Such personal fears create situations in which users refuse to cooperate with the socio-technical system. Especially, as these fears are diffuse, they are difficult to resolve. Six items measure the individual barriers. We take up the fear of losing control over data [35] by measuring the students' perception of the data stored in the background of the digital learning platforms (dlp). Having no influence on the amount and type of data storage can be intimidating [36]. With more data storage, new data analysis methods come along, and users become more traceable and might fear a lack of control [37]. Hence, students might doubt secure data handling. HEIs might potentially be able to draw conclusions on the students' individual behavior [38]. Generally, potential disruptions in the job market due to DT is a fear, not only for students. [39]. DT affects the learning environment and thus influences course delivery, learning outcomes, and job perspectives [40]. Also, DT could negatively influence instructors' job perspectives and decrease the student-to-teacher ratio.

Organizational barriers are another dimension, which are measured by using seven items. We base their construction on existing scales for change management and inertia [17]. Different HEI stakeholders might resist cultural change from traditional teaching roles or processes to new ones, even if they are the better alternatives [41]. An absence of support from the university management and a lack of strategy are often related to each other [42]. A form of management support is to solve the provision of missing resources needed to set up new structures, implement online learning concepts, or support digitally-enhanced administration as part of the DT process [22, 23].

Three items measure the impact of the dimension of technical barriers, which might hinder the DT process. These items orient towards measuring the technical interplay and integration, such as security concerns, dependence on technologies, and performance of the current infrastructure [15]. In order to partake in an online learning environment, students need suitable infrastructure in the form of continuous data connection, devices to utilize service structures, and software on their side [23]. As this is students' responsibility, it might hinder some students who don't have sufficient financial resources nor technical knowledge. We orient towards existing scales to measure the perception of the current infrastructure and security settings [42].

Four items measure the external barriers to DT. Digital learning environments combine different types of content and data. Thus, standards are needed for connection and seamless data exchange [43]. If such a standard is lacking, students might perceive it

Table 1. Questionnaire

Dimension	Items
DT Process (DT)	My university offers digital services, which support me in my studies
	My university continues to use existing methods for teaching and services
	My university is moving forward in terms of DT
	I have the impression that the university's internal processes have been digitalized
Individual Barriers (Ind.)	I am aware of the kind of (apparent) data about me that is stored when using the dlp
	I have the impression that I control the data that is stored about me
	I trust the university in handling the data I generate when using the platform
	The traceability of the data (which the lecturer can access) does not affect my use of the dlp
	I think that through the use of IT, teaching of the same quality can be done with fewer staff
	The changed form of the course harms my learning success
Organizational Barriers (Orga.)	University management supports the DT at the university
	The university has created specific jobs/projects for the DT
	At my university, we have a clear vision or DT strategy
	The learning culture at the university has not changed due to DT
	The university strives to constantly learn about and improve in how to transform digitally
	In my university, there is openness to new ideas in teaching
	I have the impression that there are not enough resources (time, money, IT staff) for the dlp
Technical Barriers (Tec.)	I have had problems with my internet connection while working with the dlp
	I have no security concerns when using the dlp
	I possess all the necessary technical means to use the dlp
External Barriers (Ext.)	I can easily integrate additional data and information into the content of the dlp
	I cannot read and edit the contents of the dlp with my standard programs
	I consider laws regulating the handling of digital products and services to be missing
	I think there are enough data protection laws that protect me in dealing with the dlp

(continued)

Table 1. (*continued*)

Dimension	Items
Missing Skills (Skills)	All in all, my IT knowledge is adequate to keep up with the changes in the course
	I have the impression that the teacher has sufficient IT skills to operate the dlp
	I don't see any advantages to the technical support provided by the dlp in the course
	I think that in the dlp, all available technical possibilities have been used
	I was integrated into the decision process about the use and scope of the dlp
	I was sufficiently trained in the use of the dlp

as a hinder. Also, standards secure access to teaching content on different devices [44]. As a lot of HEIs are public organizations, they need to have distinct security standards for data access in place, e.g., when it comes to online examinations or the electronic distribution of certificates and grades [16]. Regarding their status as a public organization, HEIs have less freedom to choose the software they use. For the formulation of the items, inspiration was taken from research on regulations [45].

Six items measure the dimension missing skills, which require special abilities to personally succeed in the DT process. Students, instructors, and administrative staff might lack sufficient IT knowledge. Thus, we survey for the perceived IT knowledge of students and teachers [46]. As internal stakeholders, students need to understand and be able to use digital concepts [47]. The understanding is essential to use available digital services and to know the right background in a digital process. Thus, items measure the perception of received training [48].

All items of the dimensions were measured on a 5-point Likert scale ranging from "I strongly agree" to "I do not agree at all". In total, participants had to indicate their perceived barriers and the DT process in 30 items. To avoid response bias, we formulated a part of the questions positively. Items on barriers could otherwise have led to a negative framing of the respondents.

5   Results

After the data collection, we analyzed and compared the data sets. To do so, we poled all items in one direction, whereby a high value corresponds to a perception of a weaker DT process or more distinctive barriers.

In the first step of the analysis, we examined the descriptive statistics. Table 2 shows that there are both similarities and differences between the pre-pandemic and intra-pandemic groups. When looking at the DT process, it is noticeable that the intra-covid group assesses the DT process worse on a mean basis. Only smaller changes, ranging

Table 2. Descriptive statistic

Dimension	Pre-pandemic						Intra-pandemic					
	N	Minimum	Maximum	Mean	Deviation	Variance	N	Minimum	Maximum	Mean	Deviation	Variance
DT	104	1.00	5.00	3.20	.86	.74	113	1.00	5.00	3.77	.90	.81
Ind.	104	1.40	5.00	3.37	.72	.52	113	1.00	4.20	3.10	.55	.31
Orga.	104	1.40	4.83	3.05	.65	.42	113	1.60	5.00	3.63	.59	.35
Tec.	104	1.00	5.00	4.02	.80	.65	113	1.00	5.00	3.53	.97	.93
Ext.	104	1.33	5.00	3.38	.80	.63	113	1.67	5.00	3.19	.67	.44
Skills	104	1.33	5.00	3.13	.77	.60	113	1.33	5.00	3.38	.77	.60

from 0.19 to 0.27, can be observed when looking at the mean values for individual and external barriers as well as missing skills. Greater differences can be seen in organizational and technical barriers, although the direction of change is different. While organizational barriers are perceived as more salient on average in the intra-covid group, technical barriers are perceived as lower.

In the next step, we examined whether the changes in student response behavior described above have an impact on the linear correlation between barriers and the DT process. The Pearson correlation coefficients in Table 3 are showing significant relationships between several barrier dimensions and the DT process in both samples. When looking at both groups in detail, however, differences become apparent. While the magnitude of the linear relationship is often comparable, differences are particularly visible in the significance and direction of effect. While no significant correlation between individual barriers and the DT process was observed in the pre-pandemic group, these dimensions show a significant weak linear correlation [52] in the intra-pandemic group. Thus, a lower degree of DT is also accompanied by more distinct individual barriers. Similarities between the two samples can be observed at the organizational barriers. Organizational barriers and the DT process show a moderate linear relationship with significant values of 0.67 in the pre-pandemic setting and 0.66 in the intra-pandemic setting. Similar to the individual barriers in the intra-pandemic setting, the Pearson correlation coefficient implies that a lower degree of DT is accompanied by more distinct organizational barriers in both samples. When comparing the samples, differences in the direction of the linear relationship can be seen for the technical barriers. The intra-covid sample shows a significant linear correlation between the two dimensions, which was expected due to the presumed effect of barriers in general. Higher levels of technical barriers are associated with lower levels of DT. Before the pandemic, however, an opposite significant correlation was observed: Higher levels of technical barriers were accompanied by higher levels of DT. A similar pattern can be observed in the case of external barriers. While a contrasting linear relationship was observed in the pre-pandemic sample, a previously suspected relationship could be observed in the intra-pandemic sample. However, a significant correlation was only found in the pre-covid sample. No significant linear relationship can be found between the external barriers and the DT process in the intra-pandemic sample. Last but not least, the missing skills show comparable results in both samples regarding the linear correlation with the DT. Both samples show a significant linear relationship, which, however, was weaker in the intra-covid sample. Pearson's correlation shows no evidence of multicollinearity among the barrier dimensions, which is important for the following regression analysis.

The Pearson correlation coefficient does not indicate the cause-effect direction. It rather shows the linear relationship between two variables. Thus, the two data sets were analyzed with a multiple regression analysis in the next step to gain a deeper understanding. To increase the comparability of the results, we chose the inclusion method for the regression analysis. Through this, we were able to investigate which barriers have a significant impact on the DT process and how the impact changed over time. As we compare the different samples with an identical model, we focus on the unstandardized regression coefficients. In addition, the different variables are measured

Table 3. Pearson correlation matrix

Dimension	Pre-pandemic						Intra-pandemic					
	DT	Ind.	Orga.	Tec.	Ext.	Skills	DT	Ind.	Orga.	Tec.	Ext.	Skills
DT	1.00						1.00					
Ind.	.15	1.00					.26**	1.00				
Orga.	.67**	.12*	1.00				.66**	.28**	1.00			
Tec.	−.19*	.35**	−.10	1.00			.21*	.24**	.36**	1.00	,	
Ext.	−.17*	.18*	0.00	.38**	1.00		.04	.082	.20*	.20*	1.00	
Skills	.48**	.28**	.44**	.07	.18*	1.00	.26**	.37**	.14	.062	−.017	1.00

*p < 0.05 significant, **p < 0.01 significant

on identical scales of measure, eliminating the need to consider standardized regression coefficients. Table 4 shows the results of the regression.

Table 4. Regression

Variable	Pre-pandemic		Intra-pandemic	
	Coefficient	Sig.	Coefficient	Sig.
Intercept	1.027	.039	-.267	.637
Ind.	.047	.612	.065	.652
Orga.	.703	.000	1.006	.000
Tec.	−.103	.226	−.028	.704
Ext.	−.205	.013	−.111	.263
Skills	.308	.001	.186	.040
N	104		113	
R^2	.545		.475	
Adjusted R^2	.521		.450	

The pre-covid model shows three significant barrier dimensions, while the intra-covid model shows only two dimensions significant at the 0.05 level. In addition, the adjusted R^2 of the pre-pandemic model shows a higher explanatory power of 0.521 compared to 0.440 in the intra-pandemic model. In both cases, the models explain the variance of the dependent variable to a satisfactory level. The models show the strongest significant influence on the DT process for the organizational barriers, with the coefficient being even higher for the intra-covid sample. The second strongest dimension is the lack of skills. Here it is apparent that the missing skills in the pre-covid model have a higher impact on the DT process than in the intra-pandemic model. For external barriers, however, only a significant influence was found in the pre-pandemic sample. Nevertheless, the effect manifests itself differently than intended. The DT process is perceived as more intensive with an increase in external barriers. This anomaly is addressed in the later discussion. Moreover, in both regressions, no significant influence of individual and technical barriers could be proven.

All in all, both similarities and differences emerge in the results of our longitudinal trend research design. In the pandemic, the DT process was perceived to be lower by the students. A stronger perception of organizational barriers was observable.

6 Discussion

In our study, the adjusted R^2 decreased over time. Thus, there are factors that are not considered in the questionnaire. The questionnaire is based on qualitative interviews conducted before the pandemic and might not have covered specific pandemic-related issues. Also, stress-related issues might overlap with the respondent's answers.

In addition, it is likely that the pandemic does not affect all students equally. Aucejo et al. [49] highlighted in a study that low-income students were more affected by the impact. Perceptions of barriers and DT may be influenced to a greater extent by students' individual circumstances.

During the pandemic, students perceive the DT process to be weaker. The perception is the result of higher digital awareness, as digitalization is often the solution to pandemic-related problems of social distancing. Regarding the correlation with other dimensions, nearly all of them decreased. In general, it seems that the weaknesses of the DT got more visible to students as a stressed DT might not deliver better results.

As for the individual barriers, the mean decreased by 0.27 only, which indicates the students perceive them to be about the same at the two different time points. The correlation with the organizational barriers and skills increased. At the same time, the correlation with technical and external barriers decreased. This indicates a closer connection between individual barriers, organizational barriers, and skills. This group of barriers could be perceived as internal factors, as students are the internal users of the HEIs' teaching offers. One important factor in this dimension is the role of trust, in which HE should serve as a role model [35]. Other authors have highlighted factors, such as individual resistance and technophobia [21], which could partly be true in this study. Although, the standard deviation is rather low in the intra-pandemic sample, indicating a rather homogenous point of view. As the surveyed students are technology-interested management students, our sample could be biased when it comes to technophobia.

Organizational barriers seem to be the key barrier in our research. The difference of the pre- and intra-pandemic mean is leaning more towards the negative side of the scale. At the same time, the standard deviation decreased a bit, which means the perception of students got more homogenous. The correlation with technical barriers and skills changed to some extent, even with a change in direction (from negative to positive). The coefficients are in both cases significant, with a huge increase in the coefficient. These results approve the results of other studies, in which the organizational factors are the major key to success or to hinder [3]. It shows that in the students' perception, this barrier got worse. Of course, HEIs had to quickly shift to online teaching, which might indicate that not all organizational processes were in place.

The mean of the technical barriers decreased rather substantially by 0.5. Important shifts in the correlation with external barriers, organizational barriers, and the DT process are visible. For the DT process and the organizational barriers, the correlation even changes from negative to positive. Even in the correlation analysis, the coefficient decreased without being significant. This could be interpreted in a way that technical barriers are not perceived as barriers anymore. Instead, the functioning of the technical side, such as the dlp, is perceived as less disturbing. By being forced to use these technologies, students probably had positive experiences. As other studies show, students are more satisfied with digitally transformed courses [50], which might also interplay with other barriers and the overall DT process.

The means of the external barriers are rather stable. The intra-covid sample shows a small decrease in the mean and the standard deviation. The correlation with skills shifts to negative, whereas the correlation with the DT process shifts to positive. In general, the values are relatively low. It might be a weak signal that skills are negatively affected by new external requirements, e.g., when new standards evolve. The influence on the DT process in the regression analysis and its significance decreased, showing less impact on the DT process. In general, students might not be involved enough in external developments or judge this barrier as less important as it is a factor that simply must be accepted, such as the General Data Protection Regulation settings.

The last dimension of missing skills is also rather stable in terms of its means and standard deviations. Thus, the general perception of this barrier is not substantially influenced by the pandemic. In correlation to other barriers, there is a decrease in relation to the DT process, the organizational barriers, and the external barriers. The coefficient in the regression analysis is decreasing and is non-significant in the intra-pandemic sample. It seems students value their skills as relatively stable but perceive them as less connected to other barriers. An increased correlation exists with individual barriers. Both dimensions could be interpreted as a personal perspective and thus perceived as rather stable.

In our study, we surveyed management students taking elective courses with an IS focus. Other studies compared the online activity level of students from different faculties. [51] In general, students are less active during the pandemic. Still, technology-related programs show a higher degree of engagement than management-related programs. Thus, we expect our respondents to be more positive towards the enforced change in the course delivery.

7 Conclusion and Limitations

Our longitudinal trend study examines the impact of the pandemic on the perceived barriers of HEIs' DT process from the perspective of management students. We aimed to determine commonalities and differences between a pre-pandemic and intra-pandemic sample based on individual, organizational, technical, and external barriers, as well as missing skills. By considering different dimensions, a socio-technical perspective was obtained.

From an overall perspective, the pre- and intra-pandemic data show many similarities. Nevertheless, they differ in the details. The lower R^2 of the regression model indicates that the DT process is influenced by additional factors than the barriers included in the questionnaire. Due to the pandemic, HEIs were forced to take steps towards DT. Thus, shortcomings in the process were highlighted. Consequently, the students in our study rate DT worse and primarily perceive organizational factors as the cause for this. For the future, however, it is unclear whether barriers have been overcome sustainably or whether they have been reinforced. A post-pandemic study could provide evidence on this.

The study and its findings should be interpreted with limitations. We surveyed students before and during the pandemic. Cohorts with different management students in the same courses with the same curricula were involved in this study. Higher reliability of results would have been expected if the same cohort of students had been captured over time. Given the high turnover of students combined with an anonymous instrument, such a research approach was unfortunately not possible. In addition, it should be considered that the courses are electives that belong to the field of information systems. This leads us to expect that we surveyed a certain kind of student audience with a basic affinity for technology regardless of the cohort. Thus, it will influence the reliability of the results in a positive way.

Also, our study shows a limited perspective on the digitalization of HEIs. As described, students are one stakeholder group. Academic faculty, administrative personnel, the general public, and the local community around HEIs have to be included in future studies, too.

To sum up, our results should still be validated by a more diversified group of students. Also, further need to pay attention to different stakeholder perspectives, e.g., by using case study approaches. Follow-up studies could pay particular attention to technology affinity but also include more in-depth sociodemographic data of students. Since the underlying model and questionnaire were developed prior to the pandemic, further qualitative research could provide additional insights that were not addressed in this questionnaire. The impact of a pandemic is complex. Its intensity can be influenced by factors such as income and living situation. These varying experiences can lead to different perceptions of the same barriers.

References

1. Fitzgerald, M., Kruschwitz, N., Bonnet, D., Welch, M.: Embracing digital technology: a new strategic imperative. MIT Sloan Manag. Rev. **55**, 1–12 (2013)
2. Castro, R.: Blended learning in higher education: trends and capabilities. Educ. Inf. Technol. **24**(4), 2523–2546 (2019). https://doi.org/10.1007/s10639-019-09886-3
3. Reid, P.: Categories for barriers to adoption of instructional technologies. Educ. Inf. Technol. **19**(2), 383–407 (2012). https://doi.org/10.1007/s10639-012-9222-z
4. Vogelsang, K., Liere-Netheler, K., Packmohr, S., Hoppe, U.: Barriers to digital transformation in manufacturing: development of a research agenda. In: Proceedings of the 52nd Hawaii International Conference on System Sciences, pp. 4937–4946 (2019)
5. Dwivedi, Y.K., et al.: Impact of COVID-19 pandemic on information management research and practice: transforming education, work and life. Int. J. Inf. Manage. **55**, 102211 (2020). https://doi.org/10.1016/j.ijinfomgt.2020.102211
6. García-Morales, V.J., Garrido-Moreno, A., Martín-Rojas, R.: The transformation of higher education after the COVID disruption: emerging challenges in an online learning scenario. Front. Psychol. (2021). https://doi.org/10.3389/fpsyg.2021.616059

7. Mishra, L., Gupta, T., Shree, A.: Online teaching-learning in higher education during lockdown period of COVID-19 pandemic. Int. J. Educ. Res. Open. **1**, 100012 (2020). https://doi.org/10.1016/j.ijedro.2020.100012

8. Marinoni, G., van't Land, H., Jensen, T.: THE IMPACT OF COVID-19 ON HIGHER EDUCATION AROUND THE WORLD. International Association of Universities, Paris (2020)

9. Vogelsang, K., Brink, H., Packmohr, S.: Measuring the barriers to the digital transformation in management courses – a mixed methods study. In: Buchmann, R.A., Polini, A., Johansson, B., Karagiannis, D. (eds.) BIR 2020. LNBIP, vol. 398, pp. 19–34. Springer, Cham (2020). https://doi.org/10.1007/978-3-030-61140-8_2

10. Hirsch-Kreinsen, H.: Digitization of industrial work: development paths and prospects. J. Labour Market Res. **49**(1), 1–14 (2016). https://doi.org/10.1007/s12651-016-0200-6

11. Friga, P.N., Bettis, R.A., Sullivan, R.S.: Changes in graduate management education and new business school strategies for the 21st century. AMLE **2**, 233–249 (2003). https://doi.org/10.5465/amle.2003.10932123

12. Löffler, A., Prifti, L., Knigge, M., Kienegger, H., Krcmar, H.: Teaching business process change in the context of the digital transformation: a review on requirements for a simulation game. Multikonferenz Wirtschaftsinformatik (MKWI) 759–770 (2018)

13. Laurell, C., Sandström, C., Eriksson, K., Nykvist, R.: Digitalization and the future of management learning: new technology as an enabler of historical, practice-oriented, and critical perspectives in management research and learning. Manage. Learn. **51**, 1350507619872912 (2019). https://doi.org/10.1177/1350507619872912

14. Vogelsang, K., Droit, A., Liere-Netheler, K.: Designing a flipped classroom course–a process model. In: Proceedings of the 14th International Conference on Wirtschaftsinformatik, pp. 345–359 (2019)

15. Whitaker, J., New, J.R., Ireland, R.D.: MOOCs and the online delivery of business education what's new? What's not? What now? AMLE. **15**, 345–365 (2016). https://doi.org/10.5465/amle.2013.0021

16. Arbaugh, J.B.: What might online delivery teach us about blended management education? Prior perspectives and future directions. J. Manag. Educ. **38**, 784–817 (2014). https://doi.org/10.1177/1052562914534244

17. Smuts, R.G., Lalitha, V.V.M., Khan, H.U.: Change management guidelines that address barriers to technology adoption in an HEI context. In: 2017 IEEE 7th International Advance Computing Conference (IACC), pp. 754–758 (2017). https://doi.org/10.1109/IACC.2017.0156

18. Irons, L.R., Keel, R., Bielema, C.L.: Blended learning and learner satisfaction: keys to user acceptance? USDLA J. **16** (2002)

19. Janson, A., Söllner, M., Bitzer, P., Leimeister, J.M.: Examining the effect of different measurements of learning success in technology-mediated learning research. In: 35th International Conference on Information Systems (ICIS), pp. 1–10 (2014)

20. Scherer, R., Siddiq, F., Tondeur, J.: The technology acceptance model (TAM): a meta-analytic structural equation modeling approach to explaining teachers' adoption of digital technology in education. Comput. Educ. **128**, 13–35 (2019). https://doi.org/10.1016/j.compedu.2018.09.009

21. Abrahams, D.A.: Technology adoption in higher education: a framework for identifying and prioritising issues and barriers to adoption of instructional technology. J. Appl. Res. High. Educ. **2**, 34–49 (2010)

22. Porter, W.W., Graham, C.R., Bodily, R.G., Sandberg, D.S.: A qualitative analysis of institutional drivers and barriers to blended learning adoption in higher education. Internet High. Educ. **28**, 17–27 (2016)

23. Al-Senaidi, S., Lin, L., Poirot, J.: Barriers to adopting technology for teaching and learning in Oman. Comput. Educ. **53**, 575–590 (2009). https://doi.org/10.1016/j.compedu.2009.03.015

24. Gregory, M.S.-J., Lodge, J.M.: Academic workload: the silent barrier to the implementation of technology-enhanced learning strategies in higher education. Distance Educ. **36**, 210–230 (2015). https://doi.org/10.1080/01587919.2015.1055056

25. Burch, Z.A., Mohammed, S.: Exploring faculty perceptions about classroom technology integration and acceptance: a literature review. Int. J. Res. Educ. Sci. **5**, 722–729 (2019)

26. Gonzalez, T., et al.: Influence of COVID-19 confinement on students' performance in higher education. PLoS ONE **15**, e0239490 (2020). https://doi.org/10.1371/journal.pone.0239490

27. Creswell, J.W.: A Concise Introduction to Mixed Methods Research. SAGE, Los Angeles (2015)

28. Petter, S.C., Gallivan, M.J.: Toward a framework for classifying and guiding mixed method research in information systems. In: Proceedings of the 37th Annual Hawaii International Conference on System Sciences, pp. 1–10 (2004). https://doi.org/10.1109/HICSS.2004.1265614

29. Fuglsang Østergaard, S., Graafland Nordlund, A.: The 4 biggest challenges to our higher education model – and what to do about them Adam. World Economic Forum, Davos (2019)

30. Elmuti, D.: Can management be taught? If so, what should management education curricula include and how should the process be approached? Manag. Decis. **42**, 439–453 (2004). https://doi.org/10.1108/00251740410523240

31. Borg, W.R., Gall, M.D.: Educational Research: An Introduction. Longman, New York (1989)

32. Adi Syani, P., Rahiem, M.D.H., Subchi, I., Suryani, R., Kurniawan, F.: COVID-19: accelerating digital transformation for university's research administration. In: 2020 8th International Conference on Cyber and IT Service Management (CITSM), Pangkal Pinang, Indonesia, pp. 1–6. IEEE (2020). https://doi.org/10.1109/CITSM50537.2020.9268913

33. Moorthy, K., et al.: Barriers of mobile commerce adoption intention: perceptions of generation X in Malaysia. J. Theor. Appl. Electron. Commer. Res. **12**, 37–53 (2017). https://doi.org/10.4067/S0718-18762017000200004

34. Klötzer, C., Pflaum, A.: Toward the development of a maturity model for digitalization within the manufacturing industry's supply chain. In: Proceedings of the 50th Hawaii International Conference on System Sciences, pp. 4210–4219 (2017). https://doi.org/10.24251/HICSS.2017.509

35. Schnackenberg, A., Tomlinson, E.: The role of transparency in the trustworthiness-trust relationship. Acad. of Mgmnt. Proc. **2012**, 15203 (2012). https://doi.org/10.5465/AMBPP.2012.15203abstract

36. Venkatesh, V., Morris, M.G., Davis, G.B., Davis, F.D.: User acceptance of information technology: toward a unified view. MIS Q. **27**, 425–478 (2003)

37. Cramer, H., et al.: The effects of transparency on trust in and acceptance of a content-based art recommender. User Model. User-Adap. Inter. **18**, 455–496 (2008). https://doi.org/10.1007/s11257-008-9051-3

38. Al-Jabri, I.M., Roztocki, N.: Adoption of ERP systems: does information transparency matter? Telematics Inform. **32**, 300–310 (2015). https://doi.org/10.1016/j.tele.2014.09.005

39. Cech, F., Tellioğlu, H.: Impact of the digital transformation: an online real-time delphi study. arXiv preprint, pp. 1–15 (2019)

40. Proserpio, L., Gioia, D.A.: Teaching the virtual generation. AMLE **6**, 69–80 (2007). https://doi.org/10.5465/amle.2007.24401703

41. Polites, G.L., Karahanna, E.: Shackled to the status quo: the inhibiting effects of incumbent system habit, switching costs, and inertia on new system acceptance. MIS Q. **36**, 21–42 (2012)

42. Bienhaus, F., Haddud, A.: Procurement 4.0: factors influencing the digitisation of procurement and supply chains. Bus. Process. Manage. J. **24**, 965–984 (2018). https://doi.org/10.1108/BPMJ-06-2017-0139.

43. Wixom, B.H., Todd, P.A.: A theoretical integration of user satisfaction and technology acceptance. Inf. Syst. Res. **16**, 85–102 (2005)
44. Piccoli, G., Rodriguez, J.A., Palese, B., Bartosiak, M.: The dark side of digital transformation: the case of information systems education. In: 38th International Conference on Information Systems, Seoul, vol. 201, pp. 1–20 (2017)
45. Ramsey, E., McCole, P.: E-business in professional SMEs: the case of New Zealand. J. Small Bus. Enterp. Dev. **12**, 528–544 (2005). https://doi.org/10.1108/14626000510628207
46. Wang, T., Jong, M.S., Towey, D.: Challenges to flipped classroom adoption in Hong Kong secondary schools: overcoming the first- and second-order barriers to change. In: 2015 IEEE International Conference on Teaching, Assessment, and Learning for Engineering (TALE), pp. 108–110 (2015). https://doi.org/10.1109/TALE.2015.7386025
47. Koehler, M.J., Mishra, P., Kereluik, K., Shin, T.S., Graham, C.R.: The technological peda-gogical content knowledge framework. In: Spector, J.M., Merrill, M.D., Elen, J., Bishop, M.J. (eds.) Handbook of Research on Educational Communications and Technology, pp. 101–111. Springer, New York (2014). https://doi.org/10.1007/978-1-4614-3185-5_9
48. Buabeng-Andoh, C.: Factors influencing teachers' adoption and integration of information and communication technology into teaching: a review of the literature. Int. J. Educ. Dev. Using Inf. Commun. Technol. **8**, 136–155 (2012)
49. Aucejo, E.M., French, J., Ugalde Araya, M.P., Zafar, B.: The impact of COVID-19 on student experiences and expectations: evidence from a survey. J. Public Econ. **191**, 104271 (2020). https://doi.org/10.1016/j.jpubeco.2020.104271
50. Arbaugh, J.B., Duray, R.: Technological and structural characteristics, student learning and satisfaction with web-based courses: an exploratory study of two on-line MBA programs. Manag. Learn. **33**, 331–347 (2002). https://doi.org/10.1177/1350507602333003
51. Aristeidou, M., Cross, S.: The impact of the Covid-19 disruption on distance learning higher education students and activities. In: 7th International Conference on Higher Education Advances (HEAd 2021). Universitat Politècnica de València (2021). https://doi.org/10.4995/HEAd21.2021.12989
52. Ratner, B.: The correlation coefficient: Its values range between +1/-1, or do they? J. Target. Meas. Anal. Mark. **17**(2), 139–142 (2009). https://doi.org/10.1057/jt.2009.5

Understanding E-Learning Styles in Distance Learning in Times of the Covid-19 Pandemic – Towards a Taxonomy

Christin Voigt$^{(\boxtimes)}$, Thuy Duong Oesterreich$^{(\boxtimes)}$, Uwe Hoppe, and Frank Teuteberg

University of Osnabrück, Katharinenstraße 1-3, 49074 Osnabrück, Germany
{christin.voigt,thuyduong.oesterreich,uwe.hoppe,
frank.teuteberg}@uni-osnabrueck.de

Abstract. The main purpose of this study is to empirically explore the e-learning behavior of university students in distance learning during the Corona-Pandemic to gain deeper insights that can help to develop more individualized e-learning practices. Based on a dataset of 164 active and former students from different study programs, universities, and semesters, we first apply factor analysis to identify 24 relevant learning factors regarding their mental progress, social aspects, and sensory perception. These factors, in turn, served as the basis for a cluster analysis, in which the students were classified into eight distinct e-learning clusters representing the taxonomy of different e-learning styles in distance learning. Based on the findings, we highlight the implications for research and practice and derived a set of seven propositions for appropriate teaching and learning strategies for distance learning. These propositions could help to address the individual digital needs of the students in a more effective manner.

Keywords: Taxonomy · Digital learning · E-learning styles · Cluster analysis · Factor analysis

1 Introduction

Since the outbreak of Covid-19, the role of digital teaching and learning has substantially increased. Educational institutions all over the world were forced to switch from face-to-face teaching to distance learning concepts overnight [1]. This quick shift has posed major challenges for both the educational institutions and learners alike. Online or digital learning refers to "learning experiences in synchronous or asynchronous environments using different devices (e.g., mobile phones, laptops, etc.) with internet access" [1]. The terms online learning or e-learning relate to a common form of distance learning that offers both learners and teachers several advantages such as enhanced flexibility and accessibility [2]. For digital teaching to be successful, both the students should have digital competences as well as the university as an institution should be well equipped digitally. Personalized teaching and learning represents one of the most challenging issues that can be facilitated by e-learning resources to create a collaborative and interactive learning environment that considers the individual needs of the students. In this

© Springer Nature Switzerland AG 2021
R. A. Buchmann et al. (Eds.): BIR 2021, LNBIP 430, pp. 19–35, 2021.
https://doi.org/10.1007/978-3-030-87205-2_2

context, it is of importance for an effective learning experience that teachers' teaching styles correspond with students' learning styles [3]. In line with the common view in the literature, we define a learning style as the individuals' attitudes and behaviors toward the preferred way of learning [4, 5]. Against this background, the main purpose of this study is to empirically examine the e-learning behavior of university students to gain deeper insights into the most prevalent learning styles that can help to develop more individualized e-learning strategies and concepts. By applying a cluster analysis, we then develop a taxonomy of e-learning styles based on the distinct styles of e-learning habits, needs and styles. These insights help us to identify appropriate teaching strategies and ways to address the individual needs of the students in a more effective manner. Therefore, the central research questions (RQ) are:

RQ1:What are the main factors and characteristics of e-learning behavior that can be integrated to develop a taxonomy of e-learning styles?

RQ2:How can these e-learning styles be addressed in distance learning through appropriate teaching strategies and concepts?

In addressing these RQs, our study contributes to the body of knowledge in the e-learning research field by proposing an empirically based taxonomy of e-learning styles that can help to improve e-learning effectiveness and e-learning experience in higher education.

2   Theoretical Background

In university research on learning theories, a distinction is made between *learning styles*, *learning strategies* and *teaching strategies* [5–7]. *Learning styles* are defined as each person's characteristics, strengths and preferences in how to take in and process information [5, 8]. This way of learning is not consistent, but varies from person to person [9]. Students can either follow a mix of learning styles or have a dominant learning style that is their personal best way to learn. Different learning styles may also be followed in different circumstances [10]. Learning styles are closely related to *learning strategies* [7]. Scarcella and Oxford define a learning strategy as the "specific actions, behaviors, steps or techniques" while learning [11]. Thereby, the learning style of a learner includes a choice of different learning strategies. The correspondence between the students' learning styles and the lecturer's teaching styles improves the interest and learning success of students, since different teaching styles are appropriate for different learning styles [3]. Therefore, teaching styles that address numerous learning styles can increase learning success and motivation [12, 13].

Extant research has found different ways to conceptualize learning styles. Vermunt analyzed the mental progress and particularly considered the processing and regulation strategies, mental models und the learning orientation. Using the Inventory of Learning Styles, he identified four different styles of learners [14–17]. Beyond this, learning styles can be classified according to the social component of learning strategies. The Grasha-Reichmann style considers three different learning styles: Students who either rely on instructions from their teacher (*Depended Learners),* or require interactions with fellow students (*Collaborative Learners*), or learn best on their own *(Independent Learner)* [18]. In addition to social distinction, learning styles can be distinguished among sensory perceptions. The *visual learning style* includes the use of pictures, images, colors, visual sights, symbols and maps [19, 20], while the *aural learning style* prefers sound and music [19, 20] and *read/write* describes a test of sensory [19]. As e-learning practices has become increasingly important, given the major role of learning practices in facilitating distance learning in times of the Covid-19 pandemic, the question of how these learning styles apply to online learning environments has become more relevant [1]. Prior attempts to examine the learning styles were primarily focused on reviews of existing learning style models and their suitability for the e-learning setting [21, 22]. Other scholars integrated selected learning styles into their concepts and prototypes of adaptive online learning environments [4, 23–25] or examine the impact of individualized and adaptive learning on the performance of e-learners [4, 23, 26]. To our knowledge, the only taxonomy of learning styles for being used in an adaptive e-learning system so far was proposed by Brown et al. (2005) who developed their taxonomy based on insights from a literature review rather than on empirical evidence. Our study is intended to close this research gap, by proposing an empirically based taxonomy of e-learning styles.

3 Research Methodology

To develop the taxonomy of e-learning styles, we follow the guidelines of Bailey [27] who defines a taxonomy as a classification of empirical entities based on a distinct set of principles, procedures and rules. As a descriptive tool, taxonomies are considered beneficial for classifying different styles of knowledge. Thereby, taxonomies reduce complexity in particular research areas, while offering researchers the possibility to identify similarities and differences among the styles for a more meaningful comparison of types and classes [27]. Overall, the development process of the taxonomy consists of three successive steps. First, we draw on the common findings of prior learning theories to identify the most important factors and characteristics of e-learning styles in distance learning, which were integrated into an online survey. Second, we collected empirical data among current and former students at German universities that serve as the basis for the taxonomy development. This survey was conducted during the Corona-Pandemic only. Third, we applied a cluster analysis as a quantitative method of classification. Since we used many items to measure the different learning theories in total, we test the construct validity using factor analysis and categories the clusters according to the factors identified. This approach of using factor scores has already been argued in previous research [28–33].

3.1 Research Design and Data Collection

The online questionnaire was developed based on the valid constructs of prior learning styles theories. For mental progress, items from Vermunt's inventory of learning styles [15–17] were integrated, while sensory perception was based on the inventory of Fleming and Mills and the typifications of Kleinginna [19, 20]. The social aspects were assessed according to the Grasha-Reichmann learning style [18]. For the collaborative learning style, two items were specifically adopted for the situation in distance learning. In addition, the students' self-perceived digital competencies, age, gender program, number of semesters, previous work experience, strongest motivation to study, preferred lecture format and preferred place of learning were surveyed. A total of 68 items were included, of which 59 items address the aforementioned learning theories. Responses were recorded on a 5-point Likert scale, with the endpoints ranging from "I completely agree" to "I do completely disagree". In the course of the evaluation, all negatively formulated items were coded positively. The survey questionnaire was provided online using LimeSurvey. After completion, the questionnaire was pre-tested with ten researchers to ensure the instrument's consistency and validity. The final survey was conducted in the digital winter-semester during the time period from September 2020 to February 2021. The participants were recruited from multiple undergraduate courses of various universities and study programs. Overall, a total of 245 students were surveyed who were either actively studying or had completed a study program at a German higher education institution and had gained experience with distance learning. Of all completed questionnaires, 164 data sets that fulfilled the conditions of completeness and accumulated experience were included into the subsequent data analysis step. The demographic details of the 164 respondents are summarized in Table 1.

Table 1. Demographic characteristics.

Characteristics	Abs.	Perc.	Characteristics	Abs.	Perc.	Characteristics	Abs.	Perc.
Gender z	78	47.5	Age <20	35	21.3	Information systems	13	7.9
Female	85	51.9	20 - 29	110	67.1	Civil service studies	5	3.1
Divers	1	0.6	30 - 39	14	8.5	Business economics	81	49.4
Total	164	100	40 - 49	1	0.6	Engineering	12	7.3
Semester 1 – 3	56	33.1	> 50	4	2.4	Teaching profession	7	4.3
4 – 5	14	8.5	Total	164	100	Medicine or health care	5	3.1
6 – 7	20	12.2	Employed and graduated students	23	14.0	Natural sciences and mathematics	6	3.7
8 – 9	16	9.8	Training position	6	3.6	Law	10	6.1
10 -11	34	20.7	Bachelor's student	91	55.5	Social sciences and humanities	16	9.8
12 -13	12	7.3	Master's student	41	25.0	Others	2	1.2
> 14	12	7.3	Others	3	1.8	Not stated	7	4.3
Total	164	100	Total	164	100	Total	164	100

Abs.: Absolut, Perc.: Percent

3.2 Data Analysis

To identify the e-learning styles in distance learning, the classifications of learning styles from the literature described above are first examined within our sample. The Cronbach's Alpha is 0.895. The factor analysis is conducted for the four categories: social aspects (independent, dependent, collaborative), sensory perception (auditory, visual, read/write) and learning styles according to Vermunt based on processing and regulation strategy, learning orientation and mental models (meaning directed, reproduction, application, undirected style). We used the Varimax rotation for factor analysis [34]. Some of the factors identified so far could be confirmed, others differed from existing theory. These new factors formed the basis of the cluster analysis using the K-Means algorithm [35]. Within the clusters, the demographic characteristics were further analyzed. In addition to the description of characteristics of each cluster, a cross-tabulation using the chi-square statistic was carried out for each cluster in order to analyse significant differences within the learning styles. Coefficients from a ten percent confidence interval upwards were supposed.

4 Taxonomy of E-Learning Styles

The statistical results are divided into factor and cluster analysis [31]. The factors form the classification according to the observed variables. In addition to the cluster analysis, the demographic characteristics of each cluster members are analyzed.

4.1 Factors of E-Learning Styles

A total of 24 relevant factors were identified that can be assigned to three different categories. Vermunt focuses on the mental progress during the learning process and considers processing strategies, regulation strategies, mental models and learning orientation. The Kaiser-Meyer-Olkin-value (KMO) for the mental progress is 0.76. The eleven identified factors including a brief description are shown in Table 2.

The second factor analysis refers to the social aspects of learning (KMO = 0.68). Prior research has shown that students learn either dependently, collaboratively or independently [18]. In the factor analysis, however, the items related to the Corona-Pandemic form a separate factor. In addition to the distinction between dependent (instructions needed) and independent (self-employed) learning styles, a mix of both styles was identified (self-reliant under instructions). The third factor analysis considers the sensory perception (cf. Table 4) during the learning progress and shows a KMO of 0.53.

The visual and auditive learning style both could be confirmed. However, instead of writing/writing, the factor analysis showed a new factor with a mix of reading and listening. Besides, the factor "writing" was determined, which involves working with text while using visual learning strategies like marking most important passages.

4.2 Clusters of E-Learning Styles

Based on the presented factors, a cluster analysis is carried out, which classifies the respondents into a total of eight clusters depending on the degree to which the factors were found in their answers. Table 5 shows the means of squares for each factor as well as the significance level and the F statistics. Apart from Factor 2, all factors are significant at least at the ten percent level of significance.

Table 2. Factor analysis: Mental progress.

Factor	Description	Items
The Practician	The courses should consist of many exercises that help students to solve practical problems. Learning means being able to apply knowledge in everyday life and the learning material is connected with practical experiences	5
The Transformer	Students transfer the learning contents into their everyday life as well as into practice and pursue not only the goals set by, but also their own. They test their learning progress using examples and problems in addition to the course material. Components are analysed stepwise	5
The Organized	Before students start learning, they figure out how to proceed best, for example using a learning plan. They are very dedicated and often do more in a course than is expected of them. They work through the lecture point by point and prefer a quiet learning without distractions	5
The Implementer	Students try to find examples of their own while learning. They search for connections between the topics covered in different chapters and try to apply the material practically	3
The Analyst	Students compare the conclusions made in the course with each other and examine whether they are based on facts from the reference book. They try to combine the different topics and find relationships in the subject matter	4
The Overstrained	Students find it difficult to determine whether they have sufficiently mastered the subject matter. They struggle to work on their own and have problems with large amounts of learning material	3
The Reproducer and Comparator	Students understand learning as a reproduction of the course facts. At the same time, however, they want to be encouraged to compare the different theories of a course with each other	2
The Memorizer	Students learn everything exactly as they find it in textbooks or scripts and memorize definitions as literally as possible	2

(*continued*)

Table 2. (*continued*)

Factor	Description	Items
The Utilitarian Differentiator	Students understand learning as examining a problem from different angles, including so far new aspects. The information should be useful either immediately or in the long-term	2
The Limiter	Students prefer compact course material that is limited to the essentials. They only use the content from the lecture and do not consult additional books	2
The Comprehensive	Students prefer to receive as much information as possible and proof their learning progress with the provided exercises and tasks within the course	2

Table 3. Factor analysis: Social aspects.

Factor	Description	Items
Collaborative	Students enjoy discussing topics with fellow students and asking for help. They also review their learning progress together with other students	4
Instructions needed	Students need structures to learn and prefer to be instructed precisely on how to solve a task	2
Missed contact during COVID19	During the Corona semester, students missed the contact with their fellow students and teachers	2
Self-reliant under instructions	Students like to shape their learning freely. At the same time, they want their teacher to motivate them and explain what is important and what is not	3
Self-employed	Students find it easy to motivate themselves to learn and to concentrate. When they have difficulties, they want their fellow learner to encourage them to find a solution on their own	3

As shown in Fig. 1, students of Cluster 1 are practical learners. Besides, they do not tend to learn with other students, but need clearer instructions from their teacher on how to solve a task. They can handle all sensory perceptions, but strongly prefer visual learning, followed by auditory learning. Cluster 1 is represented by 24% of all respondents. The two most strongly represented study motivators within this cluster include personal interest regarding the study content as well as the requirement for a profession of personal interest. Interestingly, a predominantly large number of males belongs to Cluster 1.

Table 4. Factor analysis: Sensory perception.

Factor	Description	Items
Writing	Students mark the most important passages in the textbooks and write summaries of the learning content	2
Visual	Students prefer visually oriented texts with many pictures and illustrations	3
Reading and Listening	Students prefer a textbook or script as well as an audio recording of their learning contents	2
Auditive	Students prefer to listen to music and to recite the important passages aloud while learning. At the same time, they formulate the learning content in their own words	3

Compared to the total proportion of respondents per degree program, humanities, and social sciences as well as information systems and engineering are predominantly represented. The seminar as well as exercises and tutorials are more often preferred as a lecture format. The members of the first cluster are predominantly represented in the higher semesters (6th–9th) of the bachelor's program and high technical affinity. In addition, a disproportionately large number of respondents are represented in Cluster 1 who have either gained professional experience prior to their studies or have already started their careers after completing their studies.

Table 5. Cluster analysis

Factor	ANOVA			Cluster Centers							
	Mean squares	F	Sig.	1	2	3	4	5	6	7	8
The Practician	3.66	5.17	0.00	0.68	-1.02	-0.48	-0.26	0.78	0.13	0.34	-0.22
The Transformer	1.21	1.28	0.27	0.05	-0.29	-0.69	0.00	-0.08	0.13	-0.94	1.48
The Organized	1.77	2.50	0.02	-0.02	0.13	-1.25	0.53	-0.67	-0.16	0.29	-0.80
The Implementer	1.92	2.32	0.03	0.24	0.14	-1.45	0.24	0.07	-0.51	-0.82	0.49
The Analyst	4.01	5.48	0.00	-0.35	-0.11	1.42	0.14	-1.70	0.03	2.33	-1.51
The Overstrained	3.59	6.18	0.00	-0.36	-0.01	-0.09	-0.32	1.11	0.96	-0.71	0.28
The Reproducer and Comparator	2.18	3.47	0.00	-0.13	0.10	1.80	0.13	-0.39	-0.02	-1.13	1.64
The Memorizer	2.95	3.52	0.00	-0.31	-0.56	0.35	0.20	-2.02	0.53	-0.49	-0.28
The Utilitarian Differentiator	1.64	2.00	0.07	0.20	-0.59	0.68	0.03	-0.96	0.07	0.76	1.12
The Limiter	3.28	3.84	0.00	0.01	-0.56	0.52	0.45	2.29	-0.32	-0.81	-0.38
The Comprehensive	1.68	2.04	0.06	-0.29	-0.26	0.74	0.21	1.21	-0.14	-1.43	-0.26
Collaborative	7.02	17.42	0.00	-1,2	-0.83	0.34	0.44	1.08	0.73	-1.18	0.55
Instructions needed	2.14	2.31	0.04	-0.57	0.31	0.54	0.09	-0.85	0.29	1.34	0.06
Missed Contact during COVID19	2.66	4.48	0.00	0.49	0.33	0.35	-0.33	-1,96	0.31	-0.42	0.62
Self-reliant under Instructions	5.97	10.73	0.00	0.27	-0.36	0.16	0.66	0,80	-1.01	1.57	1.48
Self-employed	2.89	3.11	0.01	-0.13	-0.43	-0.04	0.52	-1.35	-0.21	0.34	-1.83
Writing	6.91	12.86	0.00	-1.15	-0.62	0.93	0.39	0.27	0.74	-0.35	1.00
Visual	3.47	4.98	0.00	0.78	-0.47	-1.85	-0.23	-0.25	0.23	-0.26	-0.35
Reading and Listening	3.60	5.27	0.00	-0.18	-0.87	-0.51	0.09	1.87	0.20	0.32	1.86
Auditive	2.37	3.17	0.01	0.35	0.00	1.36	-0.27	1.92	-0.01	-0.69	-0.53

Cluster 2 includes 15% of all respondents. Students of the second cluster have no clear preference regarding the mental learning process. Only the rejection of application-related transfer of the learning material to everyday life and practical experience is evident. They do neglect collaborative work and need clear instructions from the professor, but tend to learn independently under instruction. Sensitively, they prefer visual and auditory learning. Cluster 2 does not include respondents from the administrative studies, law or studies for civil servants. Besides, "at home" was also most frequently chosen as the favorite place of learning. A distinction in study motivation could be identified that was significant at the five-percent significance level, which was mainly due to professional qualifications. Notably, personal interests were not mentioned by any of the respondents from Cluster 2. Furthermore, the second cluster exclusively consists of students without professional experience and their own perceived technical affinity is comparatively low within this cluster.

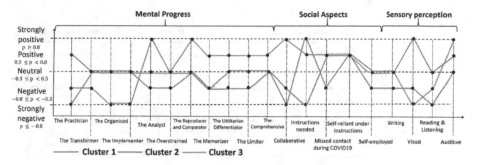

Fig. 1. E-Learning Cluster 1–3.

Cluster 3 is represented by 2% of all respondents and consists of analysts who primarily focus on comparing and weighing content. Practical application and integration play a subordinate role for this style of e-learners. As theorists, they follow a variety of learning strategies, but prefer memorization while at the same time comparing the different theories. In terms of social aspects, they are also comfortable with all factors, but prefer to work collaboratively. Cluster 3 is limited to the degree program "information systems" and a small proportion of the program "business economics". The perceived technical affinity is rated as high.

Cluster 4 e-learners tend to organize their learning process and prefer compact course materials (CM) as well as collaborative or independent learning under instructions. The preferred sensory perception is writing. All degree programs are represented in Factor 4. A high number of students in this cluster stated "to prove to others that I am capable of successfully completing a higher education program" as their motivation to study. In addition, "enjoyment of learning and studying" is mentioned most frequently within Cluster 4. Of all participants surveyed, it is the most represented cluster with a share of 29%. In addition, a large number of female respondents belong to this cluster. The own perceived technical affinity is rather high.

Cluster 5 is represented by 2% of all respondents and incorporates the highest number of practitioners and the lowest number of analysts. Members of the fifth cluster are at risk of being overwhelmed with the learning material. They tend to prefer compact course materials and like to work collaboratively or independently under instructions. However, comprehensive course materials are also strongly represented. The preferred sensory perception is the combination of reading and listening as well as auditive. Only engineering sciences are represented in Cluster 5. The preferred place of learning is in equal parts the library and at home. Students in the fifth cluster are either at the end of their master's degree or have already completed it.

Members of Cluster 6 also tend to be overwhelmed by the learning material. They learn by mere memorization and prefer collaborative learning and need clear instructions under which they also tend to be self-reliant. Without instructions, however, they are not very self-employed. In Cluster 6, solely respondents from the economics are represented. E-learning style 6 is represented by 2% of respondents, although a disproportionately large number of economists were surveyed. Both "at home" and "in learning rooms" were named as the preferred place of learning (Fig. 2).

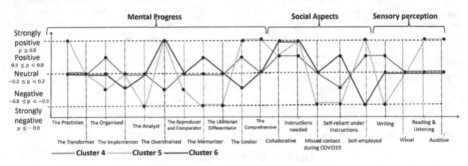

Fig. 2. E-Learning Cluster 4–6.

Learners in the seventh cluster are analysts. They are least inclined to be overwhelmed with their studies. During the Covid-19 pandemic, they missed contact with students and professors, yet not for learning, as they neither need clear instructions nor learn collaboratively. Moreover, they are very self-employed. Many business and economics students are represented in Cluster 7. In addition, all degree programs except engineering, medicine and natural sciences are represented. Overall, 22% of all respondents follow Cluster 7. A disproportionately large number of respondents gave the answer "prerequisite for a profession in which I am financially well secured" as their motivation to study. Learning takes place disproportionately often in public study rooms and in the library. Moreover, people in the seventh cluster assess their own technical affinity as high comparatively often (Fig. 3).

Members of Cluster 8 tend to be practitioners and transformers who intend to transfer or implement the learning content into their everyday life. Furthermore, they learn mostly by reproducing and comparing the content. They also aim for usefulness of the learning content, attempt to consider it from differentiated perspectives and learn independently. Furthermore, they clearly prefer the sensory style "reading and listening". Cluster 8 is represented by medical students and a small proportion of law students. It is represented by 2% of all respondents, but it is worth mentioning that medical scientists

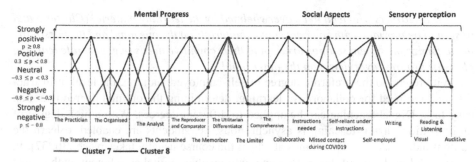

Fig. 3. E-Learning Cluster 7–8.

were underrepresented in relation to other study programs. The preferred lecture format is "lectures", and the preferred place of learning is "at home" and "in the library". With a ten percent significance, members of the eighth cluster consider themselves to have no technical affinity.

5 Implications

In line with prior research, the results of our study have shown that the e-learning behavior of students are varying greatly according to their preferences and needs [5, 8, 23]. As indicated by the findings of the factor and cluster analysis, eight different e-learning styles could be identified based on the individual mental progress, social aspects and sensory perception of the learners. Since each e-learning style differs in the learners' personal interests and motivation while revealing specific needs regarding the format of courses and learning materials, the desired degree of instruction as well as the level of collaborative learning and social contacts, the e-learning strategies and concepts should be designed to address these needs. Prior research has shown that a correspondence between learning styles and an adaptive learning environment increases learning success and effectiveness while making learning more enjoyable [4, 23, 26]. Adopting an "one size fits all" mentality while neglecting the individual e-learning styles may have negative impacts on the learners' satisfaction and performance [24]. Thus, different teaching styles are necessary for the different styles of e-learners in distance learning to optimize their learning performance [4, 23]. In particular, teaching strategies and electronic media should be individualized [5].

For online learning to succeed, teaching strategies are required that enable students to learn in an effective manner. As acknowledged by extant research, there is not only one appropriate teaching strategy for a particular learning style, but rather several teaching strategies with a set of beneficial electronic media and materials for particular e-learning styles [5]. Based on the learning behavior of the different e-Learning styles, we make the following propositions concerning the content and materials:

P1: For individuals with a high preference for practical references (e-learning style Cluster 1, 5 and 7), the learning content should be designed to include practical exercises and case studies (e.g. as online tests) through which they can actively apply their knowledge and solve problems.

P2: E-learners who tend to be overwhelmed by the learning material (Cluster 5 and 6) should be provided with compact course materials on the one hand and the possibility to customize the system settings to adapt to their cognitive needs (e.g. generate learning plans or progress overviews) on the other hand.

With regard to the preferred degree of self-employed learning, the level of instruction and the need for collaborative learning, determining the optimal degree is not an easy task. Digital teaching can be applied to different degrees. With a high degree of digitalization, digital teaching is asynchronous, i.e. students are enabled to learn the content independently of the times at which the course usually takes place [36, 37]. In contrast, a low degree of digital teaching only involves digital enrichment, with course content being partly provided digitally and organizational content being supported via a digital learning management system. Thus, in learning environments with a low degree of digitization, the teaching takes place synchronously with the focus on face-to-face teaching in the lecture hall or the synchronous virtual conference [36, 37], whereas students basically need to learn self-employed in highly digitized environments because there is no interaction with the teacher. Regardless of the degree of digitization, anecdotal evidence has shown that the individualization of pedagogical procedures can enhance the learning effectiveness of students, since learners can differ in their individual characteristics, such as educational background, and their individual needs [38]. In previous research, neither a very high nor a very low degree of digitalization has proven to increase learning success. Instead, a medium level of digital teaching with partially synchronous teaching elements which combines the advantages of digital and face-to-face teaching has been shown to be best practice [36, 37]. A past study that compared self-paced and instructor–student interactive tools, the two most common e-learning tools, found that instructor–student interactive tools perform better than self-paced tools when it comes to system quality, perceived usefulness, satisfaction, and learning outcome [38]. The reason for this can be found in the fact that instructor–student interactive tools enable more natural interaction and communication which is perceived more conducive to effective learning [38]. Besides, instructor–student interactive tools allow instructors to communicate with their students in natural language and to provide social support as well as individualized feedback that is not possible in self-paced learning environments [38]. This is especially important in distance learning where online learning is characterized by a high extent of aloneness and anonymity [39], which may be counter to the nature of some e-learners who prefer personal interaction and collaborative learning, as also revealed by the results of our study. However, face-to-face learning can be implemented in online learning environments only to a limited extent [39], e.g. via video-conferencing and thus is not fully replicable in online environments. Hence, the following proposition should be considered:

P3: Synchronous face-to-face learning settings in videoconferences should be used to enrich the learning environment to foster collaborative learning for students with a strong need for collaborative learning (Cluster 3–6 and 8).

Given the differing e-learning styles found in this study as well as the anecdotal evidence in the literature, we additionally make the following proposition regarding the level of synchronous and asynchronous teaching:

P4: An interactive learning environment with a medium level of digital teaching (self-paced learning) and partially synchronous teaching elements which combines the advantages of digital and face-to-face teaching (instructor–student interactive tools) is recommended in distance learning.

Similar to the blended learning concept, simple learning objectives such as remembering or understanding the content are outsourced to the digital space. This allows the free time in the lecture hall or in the synchronous video conference to be used for interactive and innovative activities, whereby knowledge is analyzed, applied and used with the support of the teacher and fellow students [40–42]. However, an important prerequisite for a medium level of digital teaching is that the students engage with the digital content in advance [43, 44], which in turn places more emphasis on the role of self-directed learning in such learning scenarios. For self-directed learning to be successful, learning management systems are necessary that enable students to learn effectively. Learning management systems have often been criticized for their static nature with a limited extent of adaptivity [26]. The provision of an adaptive online learning environment, as is described in prior studies [4, 23–25], would help to individualize the online learning environment to the learners' needs by automatically selecting the most appropriate learning formats, learning objects and features in accordance to the learning styles. This is especially important given the differing preferred sensory perceptions of the e-learning styles. With recent advances in artificial intelligence (AI), for example, adaptive e-learning environment can be developed to integrate interactive user-interfaces that are close to face-to-face learning settings [22]. Thus, we make the following proposition with respect to the level of adaptive learning:

P5: An adaptive e-learning environment should be used that offers students and instructors the possibility to integrate different formats of learning materials, objects, media, and features that consider the individual e-learning styles of students.

This is especially important given the fact that e-learners differ substantially with respect to their perceived degree of technical affinity and their preferred sensory perceptions. The following should be considered:

P6: The e-learning environment and tools should be designed to be user-friendly and easy to use to consider the differing degrees of technical affinity of their users.

However, two major preconditions for such an adaptive system should be mentioned in this context. The first important precondition is the prediction of the students' learning style in each course prior to preparing the course structure, materials and content, e.g. at the beginning of each semester or study program [5, 22]. AI-based techniques can help to facilitate this step based on the students' learning history [21, 45, 46]. Second, adaptive learning environments also require the provision of the learning materials in all necessary variants and formats which would take a substantial amount of time for preparing and creating the learning contents [24]. Although both issues have shown to be possible in previous studies, in daily practice it remains a challenging task to detect the students' e-learning styles in all courses and prepare the course materials in all possible variants. Rather, we propose:

P7: If possible, a trade-off should be achieved between the variety of choice and preparation time to create an online learning environment that offers students of all e-learning styles the possibility to flexibly chose the formats that are most conducive to their ability, preferences and needs.

Given the implications presented above, the findings of our study contribute to the body of knowledge in the e-learning research field by proposing an empirically based taxonomy of e-learning styles that can help to improve learning effectiveness and learning experience in e-learning settings not only in higher education, but also in organizational learning. For designers of online learning environments and e-learning applications, the taxonomy of the e-learning styles as well as the set of propositions can serve as a helpful guideline to develop adaptive learning systems that consider the e-learning styles of the students. In a similar manner, for organizations wishing to develop and introduce e-learning practices, the proposed taxonomy of the e-learning styles can help to integrate the e-learning styles of their employees into their e-learning environments to provide more effective training and education.

6 Conclusion

Given the major role of e-learning practices, both in facilitating distance learning and in the future of higher education, more individualized e-learning strategies and concepts are required that consider the individual needs of the students. To address this issue, the main purpose of this study is to gain deeper insights into the most prevalent e-learning styles of university students that can help to develop more individualized e-learning strategies and concepts. By means of a factor analysis, we identified 24 relevant factors that can be assigned to three different categories. These factors, in turn, served as the basis for the cluster analysis. Based on the findings, we provided the implications for research and practice and suggested a set of seven propositions.

As with any empirical research, our study is subject to several limitations that may affect the common claims for validity, objectivity and generalizability in social research [47]. The first limitation is concerned with the focus of this study on digital learning in higher education which does not consider digital learning practices in primary and secondary education. Although this specific focus is necessary to gain deeper insights into the specific learning conditions in higher educations, a future comparison across different educational levels would help to advance our understanding of different e-learning styles and their evolvement over time. Furthermore, within the cluster analysis, no explanatory value was given for one factor "The Transformer". Another limitation relates to the underlying data sample. As revealed by the demographic data, nearly 70% of the respondents were students of the business economics (50.9%), information systems (8.2%) and other study programs of the social sciences (8.2%). Thus, the findings include the perceptions of other student groups from other semesters, such as those from the natural sciences, linguistics, or legal sciences, only to a limited extent. Besides, the data were collected in distance learning of 2020/21 and thus only reflect the students' perception at one single point of time. A worthwhile avenue for future research is to extend the scope of this study to additionally include the perceptions of student groups from other disciplines and other semesters to receive a more comprehensive and representative picture of e-learning styles in distance learning. As a final note, the proposed taxonomy of e-learning styles may be incomplete given the aforementioned focus. According to

Bailey [27], a taxonomy should consist of exhaustive and mutually exclusive classes, a requirement that our taxonomy only partially satisfies given the aforementioned limitations. In view of the ongoing dynamic changes in higher education, the taxonomy should regularly be verified and adapted.

References

1. Dhawan, S.: Online learning: a panacea in the time of COVID-19 crisis. J. Educ. Technol. Syst. **49**(1), 5–22 (2020). 004723952093401 8
2. Al-Azawei, A., Parslow, P., Lundqvist, K.: Investigating the effect of learning styles in a blended e-learning system: an extension of the technology acceptance model (TAM). Aust. J. Educ. Technol. **33**, 2 (2017)
3. Borg, M.O., Shapiro, S.L.: Personality type and student performance in principles of economics. J. Econ. Educ. **27**, 3–25 (1996)
4. El Bachari, E., Abelwahed, E.H., El Adnani, M.: An adaptive teaching strategy model in e-learning using learners' preference: LearnFit framework. Int. J. Web Sci. **3**(1), 257–274 (2012)
5. Franzoni, A.L., Assar, S., Defude, B., Rojas, J.: Student learning styles adaptation method based on teaching strategies and electronic media. In: Presented at the 2008 Eighth IEEE International Conference on Advanced Learning Technologies (2008)
6. Helmke, A., Krapp, A.: Lehren und Lernen in der Hochschule. Einführung in den Thementeil. Zeitschrift für Pädagogik **45**, 19–24 (1999)
7. Xu, W.: Learning styles and their implications in learning and teaching. Theory Pract. Lang. Stud. **1**, 413–416 (2011)
8. Felder, R.M., Silverman, L.K.: Learning and teaching styles in engineering education. Eng. Educ. **78**, 674–681 (1988)
9. Sewall, T.J.: The Measurement of Learning Style: A Critique of Four Assessment Tools (1986)
10. Faria, A.R., Almeida, A., Martins, C., Gonçalves, R., Figueiredo, L.: Personality traits, learning preferences and emotions. In: Presented at the Proceedings of the Eighth International C* Conference on Computer Science and& Software Engineering (2015)
11. Scarcella, R.C., Oxford, R.L.: The Tapestry of Language Learning: The individual in the Communicative Classroom. Heinle & Heinle, Boston (1992)
12. Bartlett, R.L.: Discovering diversity in introductory economics. J. Econ. Perspect. **10**, 141–153 (1996)
13. Claxton, C.S., Murrell, P.H.: Learning Styles: Implications for Improving Educational Practices. ASHE-ERIC Higher Education Report No. 4, 1987. ERIC (1987)
14. Vermunt, J.: Leerstijlen en sturen van leerprocessen in het hoger onderwijs: naar procesgerichte instructie in zelfstandig denken. (1992)
15. Vermunt, J.D.: Inventory of learning styles (ILS). Tilburg University, Department of Educational Psychology, Tilburg, The Netherlands (1994)
16. Vermunt, J.D.: Metacognitive, cognitive and affective aspects of learning styles and strategies: a phenomenographic analysis. High. Educ. **31**, 25–50 (1996)
17. Vermunt, J.D.: The regulation of constructive learning processes. Br. J. Educ. Psychol. **68**, 149–171 (1998)
18. Riechmann, S.W., Grasha, A.F.: A rational approach to developing and assessing the construct validity of a student learning style scales instrument. J. Psychol. **87**, 213–223 (1974)
19. Fleming, N.D., Mills, C.: Not another inventory, rather a catalyst for reflection. Improve Acad. **11**, 137–155 (1992)

20. Kleinginna, P.R., Kleinginna, A.M.: A categorized list of emotion definitions, with suggestions for a consensual definition. Motiv. Emot. **5**, 345–379 (1981)
21. Jegatha Deborah, L., Baskaran, R., Kannan, A.: Learning styles assessment and theoretical origin in an E-learning scenario: a survey. Artif. Intell. Rev. **42**(4), 801–819 (2012). https://doi.org/10.1007/s10462-012-9344-0
22. Fatahi, S., Moradi, H., Kashani-Vahid, L.: A survey of personality and learning styles models applied in virtual environments with emphasis on e-learning environments. Artif. Intell. Rev. **46**(3), 413–429 (2016). https://doi.org/10.1007/s10462-016-9469-7
23. Abdullah, M., Daffa, W.H., Bashmail, R.M., Alzahrani, M., Sadik, M.: The impact of learning styles on Learner's performance in E-learning environment. Int. J. Adv. Comput. Sci. Appl. **6**, 24–31 (2015)
24. Brown, E., Cristea, A., Stewart, C., Brailsford, T.: Patterns in authoring of adaptive educational hypermedia: a taxonomy of learning styles. J. Educ. Technol. Soc. **8**, 77–90 (2005)
25. El-Bakry, H.M., Saleh, A.A., Asfour, T.T., Mastorakis, N.: A new adaptive e-learning model based on learner's styles. In: Presented at the Proceedings of 13th WSEAS International Conference on Mathematical and Computational Methods in Science and Engineering (MACMESE 2011). Catania, Sicily, Italy (2011)
26. Graf, S., Kinshuk, K.: Providing adaptive courses in learning management systems with respect to learning styles. In: Presented at the E-Learn: World Conference on E-Learning in Corporate, Government, Healthcare, and Higher Education (2007)
27. Bailey, K.D.: Typologies and Taxonomies: An Introduction to Classification Techniques. SAGE, Thousand Oaks (1994)
28. Cattell, R.B.: A biometrics invited paper. Factor analysis: an introduction to essentials I. The purpose and underlying models. Biometrics **21**, 190–215 (1965)
29. Christenson, A.: The requirements of factor analysis. Mankind **8**, 307–309 (1972)
30. Christenson, A.L., Read, D.W.: Numerical taxonomy, R-mode factor analysis, and archaeological classification. Am. Antiquity **42**(2), 163–179 (1977)
31. Gorman, B.S., Primavera, L.H.: The complementary use of cluster and factor analysis methods. J. Exp. Educ. **51**, 165–168 (1983)
32. Sneath, P.H., Sokal, R.R.: Numerical taxonomy. The principles and practice of numerical classification (1973)
33. Zemsky, R.: Fortran Programming for the Behavioral Sciences (1968)
34. Kaiser, H.F.: The varimax criterion for analytic rotation in factor analysis. Psychometrika **23**, 187 (1958)
35. Jain, A.K., Dubes, R.C.: Algorithms for Clustering Data. Prentice-Hall Inc., Hoboken (1988)
36. Arnold, P., Kilian, L., Thillosen, A., Zimmer, G.M.: Handbuch E-Learning: Lehren und Lernen mit digitalen Medien. UTB (2018)
37. Pfeiffer-Bohnen, F.: Vom Lehren zum Lernen: Digitale Angebote in universitären Lehrveranstaltungen. Walter de Gruyter GmbH & Co. KG (2017)
38. Hsieh, P.-A.J., Cho, V.: Comparing e-Learning tools' success: the case of instructor–student interactive vs. self-paced tools. Comput. Educ. **57**, 2025–2038 (2011)
39. Hamilton, C., Paek, L.: From blended to e-learning: Evaluating our teaching strategies. ASp. la revue du GERAS, pp. 47–64 (2020)
40. Bergmann, J., Sams, A.: Flip Your Classroom: Reach Every Student in Every Class Every Day. International Society for Technology in Education, Alexandria (2012)
41. Handke, J., Sperl, A. (eds.): Das Inverted Classroom Model: Begleitband zur ersten deutschen ICM-Konferenz. Oldenbourg, München (2012)
42. Krathwohl, D.R.: A revision of Bloom's taxonomy: an overview. Theory Pract. **41**, 212–218 (2002)
43. Handke, J.: Handbuch Hochschullehre Digital: Leitfaden für eine moderne und mediengerechte Lehre. Tectum Wissenschaftsverlag (2020)

44. Kauffeld, S., Othmer, J. (eds.): Handbuch Innovative Lehre. Springer, Wiesbaden (2019). https://doi.org/10.1007/978-3-658-22797-5
45. Kolekar, S.V., Pai, R.M., Manohara Pai, M.M.: Prediction of learner's profile based on learning styles in adaptive e-learning system. Int. J. Emerg. Technol. Learn. **12**, 31 (2017)
46. Villaverde, J.E., Godoy, D., Amandi, A.: Learning styles' recognition in e-learning environments with feed-forward neural networks. J. Comput. Assist. Learn. **22**, 197–206 (2006)
47. Shipman, M.D.: The Limitations of Social Research. Routledge, Milton Park (2014)

Quality of OER from the Perspective of Lecturers – Online Survey of Quality Criteria for Quality Assurance

Carla Reinken(✉) , Paul Greiff , Nicole Draxler-Weber , and Uwe Hoppe

Osnabrück University, Katharinenstr. 1, 49074 Osnabrück, Germany
{carla.reinken,paul.greiff,nicole-draxler-weber,
uwe.hoppe}@uni-osnabrueck.de

Abstract. The trend of digitization at universities has increased as a result of the Corona pandemic. Universities and teachers were forced to undertake a digital restructuring in a short period of time. Teaching had to be digitized, so interest in digital educational materials continued to grow. Open Educational Resources (OER) offer great potential in this context due to their digital format and usability across universities. When using as well as creating such free educational materials, lecturers have to check the quality of the OER due to individual quality requirements. For this reason, special attention should be paid to how quality assurance can be measured. There are already numerous models for quality assurance, which differ in terms of their complexity and level of detail. Zawacki-Richter and Mayrberger have conducted a comparison of these models and developed an adapted model based on their findings. By including lecturers, as a relevant stakeholder group, this model will be verified. Lecturers can be seen as exporters or importers of OER. Whether something is high quality depends on the needs of each user, their tasks and processes, so quality requirements can vary with the respective role. The aim is to investigate which quality criteria lecturers think OER should fulfill in order to be of high quality. For this purpose, an embedded mixed method (EMM) survey of lecturers was conducted, in which the different roles of a lecturer were also distinguished.

Keywords: Open Educational Resources · Quality criteria · Quality measurement · Higher education · Digitalization

1 Introduction

Learning with technologies has become a global phenomenon long ago [1]. The Corona pandemic has increased the trend of digitization at universities. In a relatively short period of time, universities were forced to act digitally and, accordingly, teachers had to digitize their teaching. Lecturers had the opportunity to put their materials into a digital format themselves or to make use of existing digital educational materials. The possibility of using existing digital materials would save lecturers time and reduce costs. Open Educational Resources (OER) provide such a digital format and can be used

© Springer Nature Switzerland AG 2021
R. A. Buchmann et al. (Eds.): BIR 2021, LNBIP 430, pp. 36–50, 2021.
https://doi.org/10.1007/978-3-030-87205-2_3

across universities, which leads to an even higher interest in OER [2]. UNESCO defines OER as digital resources for teaching, learning and research in any medium [3]. One important characteristic is the free access and use, adaptation, and redistribution by others without or with minor restrictions [4]. Lecturers can take on different roles when using OER, which will be discussed in more detail below. A distinction can be made between importer and exporter. When using OER as an importer, the quality must be checked by the lecturer, since they were created by another person. Lecturers have individual quality requirements and do not know the quality requirements of other creators. As exporter of OER they will not share their materials if they do not meet their own quality standards. This raises the question of how to measure quality to increase quality assurance and shows the relevance of taking a closer look at this topic.

Various approaches to quality assurance already exist, differing in their complexity, level of detail, and practicality. The question of quality in teaching refers, among other things, to the learning materials, whereby the increasing digitization also brings the technical and media-didactic quality criteria of OER into focus [5]. Since the lecturers themselves decide whether the external materials should be included in their own teaching or whether their own teaching content should be shared with others, the lecturers' perspective is crucial in this survey. When considering this stakeholder group, it is important to distinguish between the different roles as exporter and importer in relation to OER as already mentioned earlier. On the one hand, lecturers can incorporate OER in their own teaching and import them accordingly. On the other hand, lecturers can publish their teaching materials as OER and make them available to others, thus acting as exporters. These two roles are characterized by different approaches, which may be accompanied by different quality requirements. Therefore, it will be investigated which quality criteria OER have to fulfill in order to be published or used in own teaching. For this purpose, an embedded mixed method survey of teachers who have already created OER and made them freely available in portals, as well as of teachers who have integrated existing teaching materials into their own teaching, will be used. In the questionnaire, both the general conceptual understanding of quality in relation to OER and the concrete quality criteria of OER are queried. The quality criteria were derived based on Zawacki-Richter's and Mayrberger's [5] holistic quality assurance approach, which was developed through a literature review and a detailed comparison of well-established OER-related quality management models. In this paper, however, the quality assurance approach is to be differentiated from the following consideration of quality criteria, which are merely a tool for achieving higher quality assurance and does not represent a complete quality assurance process. Since the models considered in the work of Zawacki-Richter and Mayrberger do not have an empirical survey with lecturers, this survey represents an important benefit by including this crucial stakeholder group. In addition, the inclusion of lecturers enables a user-related weighting of the criteria, so that an evaluation of the quality criteria is possible. Finally, missing and redundant criteria for quality assurance can be identified through the results of the survey. The aim of the study is to uncover relevant quality criteria of OER based on the survey results of lecturers and to develop an adapted approach for quality measurement on an empirical basis. In this context, the following research questions arise:

1.1) What characterizes high-quality OER from the perspective of teachers?

2.1) Which quality features of existing models are perceived as particularly relevant by experts?

2.2) Which quality requirements perceived as relevant are not considered in the model? Which quality criteria from the model are redundant?

2 Theoretical Foundation

2.1 Open Educational Resources

Around the world, the Internet and other digital technologies have been used for decades to develop and distribute teaching and learning content in higher education institutions [2]. OER have gained increased attention in recent years due to their potential and promise to transcend educational demographic, economic, and geographic boundaries and promote lifelong as well as personalized learning [6]. The term Open Educational Resource was first introduced at a conference hosted by UNESCO in 2000 and was promoted in the context of providing free access to educational resources on a global scale. A very commonly used definition of OER is, "digitized materials offered freely and openly for educators, students and self-learners to use and reuse for teaching, learning and research" [7]. Since 2007, the European Commission has also been funding projects on OER (e.g. Open eLearning Content Observatory Services, OLCOS). OER are thus in the tradition of the idea of education as a common good [5]. They have emerged as a concept with great potential for supporting change in education. A key advantage is that free educational materials can be shared very easily over the Internet [8]. Important to note is that there is only one key differentiator between an OER and any other digital educational resource: its license. Thus, an OER is simply an educational resource that includes a license that facilitates reuse, modification and possibly adaptation without the need to first obtain permission from the copyright holder [8].

With the establishment of e-learning and digital learning platforms at universities, for example, there is a growing awareness of the potential and the problem areas arising from digitization in university teaching. The creation and distribution of digital materials are part of lecturers' routines. These activities take place especially in blended learning and e-learning formats, but also in traditional presence teaching [9]. In this context, it should be noted that lecturers can take on different roles and accordingly have different quality requirements for OER. On the one hand, the lecturer can act as an importer. Here, the lecturer integrates OER into his or her own teaching and must check whether the individual quality requirements are met. No lecturers want to implement OER with inferior quality into their own teaching. On the other hand, a lecturer can act as an exporter. In this case, the lecturer's own materials are made available to others in portals or repositories. In order not to be exposed, compliance with quality standards and quality checks are essential here. In both cases, quality criteria can support as a measurement of quality and increase quality assurance.Although OER are very high on the social and inclusion policy agenda and supported by many stakeholders in education, their use in higher education has not yet reached a critical threshold [10]. Possible reasons for the low usage are that OER are unknown to many lecturers, they do not know where OER can be found and how to assess the quality of the content if the OER have been created by others.

2.2 Quality

There is plenty of research in Business Administration to quality and processes like quality assurance or the encompassing quality management. The main reason is, that the quality of products and services is important for a company to compete successfully in the market [11]. In general, quality is the degree to which characteristics of an object conform to requirements [12]. Requirements are context specific and have to be considered in the respective environment. Therefore, quality management systems have become established to ensure the respective requirements. Quality management developed over several evolution phases, from simple quality control to quality assurance up to quality management systems. In contrast to quality assurance, which takes preventive measures to eliminate negative influences on the end product, quality management is a planning-conceptual design approach that takes the entire organization into account. Quality assurance is considered to be a part of quality management, aimed at generating confidence that quality requirements are based on the defined quality criteria [11]. One of the most common quality management frameworks is the ISO9000 series with focus inter alia on customer satisfaction, leadership, process management, continuous improvement and supplier relationships [13].

However, quality management has found its way into the educational landscape [14]. In the higher education system many various quality-related factors can be defined, like quality of entering students, equipment and support services, teaching quality or quality of research [15]. Even though the development of future professionals depends on the quality of education [12], the quality management systems in higher education are criticized for disregarding quality of education and putting quality for accountability purposes first [14]. The dynamic of new developments poses new challenges, because "[…] higher education has entered a new environment in which quality plays an increasingly important role" [15]. OER create new challenges regarding to quality assurance of resources, which are used in the teaching process. In response to the growing number of educational materials available on the Internet, identifying high-quality resources becomes more challenging than ever before [16]. Therefore, various quality assurance approaches for OER have been developed in the last years, which were identified and compared by different authors. The analyses show differences in depth of detail and complexity of the approaches [5, 13]. Zawacki-Richter and Mayrberger [5] identified eight approaches[1] based on a literature review and analyzed them concerning methodology, content aspects, context of application, scientific merit and existing handouts. According to the analysis, the authors distinguish between two groups of quality assurance approaches for OER: lists of criteria and scoring models, but none of them included a weighting of individual criteria. Within both groups the range extends from eight to 42 criteria. The meta-analysis revealed 168 quality criteria in total and the authors developed a synoptic overview that considers all of the identified criteria. In this way, all criteria identified in the models studied are taken into account. For this reason, Zawacki-Richter

---

[1] Learning Object Review (LORI), MERLOT Rubric, Framework for Assessing Fitness for Purpose in OER, OER Rubric (Achieve Organization), Learning Object Evaluation Instrument (LOEI), Learning Objects Quality Evaluation Model (eQNet), Rubric to Evaluate Learner Generated Content (LGC), Rubric for Selecting Inquiry-Based Activities.

Table 1. Synoptic overview according to Zawacki-Richter and Mayrberger [5]

Main quality category	Sub category	Main criteria	Sub criteria
Technological criteria	Usability	Structure, navigation, orientation	
		Interactivity	
		Design	
		Readability	
	Accessibility	Reliability	
		Interoperability and compatibility	
		Accessibility to persons with disabilities	
	Reusability	International technical standards	
Pedagogical-didactic criteria	Content	Accuracy	Correctness
			Completeness
			Actuality
		Adequacy	Target group
			Scope
		Intelligibility	Intelligibility
		Significance	Significance
			Scientific Basis
		Coherence	Coherence & Conclusiveness
		Reusability	Reusability
	Learning Design	Target group orientation	
		Learning goals and alignment	
		Engagement and motivation	
		Collaboration, communication and cooperation	
		Application and transfer	
		Assistance and support	
		Orientation towards education standards	
	Assessment	Assessment	
		Feedback	
Private international law criteria	License		

and Mayrberger's model will be considered in the following. The overview is listed in Table 1.

For the classification of the 168 criteria, Zawacki-Richter and Mayrberger used the model according to Kurilovas et al. as a basis, which was developed within the framework of the eQNet Quality Network for European Learning Resource Exchange. Thus, an orientation was made according to technological criteria, pedagogical-didactic criteria and private international law criteria, which are listed in the "main quality category" column. The designation of the four columns in Table 1 was made as an addition by the authors of this paper to provide a better understanding of the different columns of the criteria model. The main categories can be hierarchically subdivided into further sub categories (see Table 1 "sub category"), which provide information about what is meant under the main categories. Technological criteria refer to the usability of the OER, their accessibility and reusability. Pedagogical-didactic criteria were the most mentioned in the analyzed approaches and deal with the content quality of the learning materials, the didactical design and the intended forms of assessment. Private international law criteria highlight the importance of open licenses for OER. The subcategories are assigned the quality criteria ("main criteria") that will be examined in the further course. Among the pedagogical criteria quality of content awarded great importance, wherefore this dimension is also divided into a second criteria level ("sub criteria") [5].

3 Method

This paper addresses the quality requirements of OER from the perspective of lecturers as users and creators. For this purpose, an online questionnaire was developed in order to answer the aforementioned research questions as comprehensively as possible. For this reason, an embedded mixed method approach was chosen, combining both quantitative and qualitative elements [17]. The purpose of the embedded design is to have quantitative and qualitative data collected simultaneously or sequentially. Here, one form of data has a supporting role for the other form of data. The supporting data can be either qualitative or quantitative [18]. In this case the quantitative research design dominates, which is complemented by qualitative data. Therefore, the questionnaire consists of both open and closed items on the topic quality of OER. For the questions about the quality of OER II (closed items), the items were modeled according to the main criteria and sub criteria from the model of Zawacki-Richter and Mayrberger [5]. These items include questions with different answer options e.g. via dropdown selection menu and questions with a four-point importance scale in terms of a Likert-scale. In the last case, the answer options rank from "very important" to "not important". Here, an even number of answer options was deliberately given in order to prevent a central tendency bias in the sense of a forced choice scale [19]. The target group of this survey were lecturers who had already created or used OER. These were identified with the help of an online search on seven major German OER repositories (e.g. Twillo, HOOU, OpenRUB). Since this survey was aimed exclusively at German participants, the questionnaire was written accordingly in German. Before sending the questionnaire to the target group a pretest was conducted with ten researchers to check and adjust comprehensibility and wording of the items. The survey starts with a welcome text, where the purpose of the survey is explained and a definition of OER in accordance with UNESCO is given for a common understanding [3]. The questionnaire begins with two open-ended questions about quality in general, followed by a separate section of closed-ended questions about the quality of OER specifically. At the end, some demographic data is collected (see Table 2).

The survey took on average about 12 min to complete, including reading the welcome text and instructions on how to complete it. It was created using the online survey tool LimeSurvey (www.limesurvey.org) and the questionnaire was sent to the respondents

Table 2. Overview of the questionnaire structure

Question group	Question	Answer type
Quality of OER I	What characterizes good quality OER for you in general?	Free text answer
	What quality requirements do you personally expect from OER?	Free text answer
Quality of OER II	How important are the following technical quality criteria to you with regard to OER?*	Matrix question, eight individual items, four-point importance scale
	How important to you are the following pedagogical-didactic quality criteria with regard to OER?*	Matrix question, eleven individual items, four-point importance scale
	How important to you are the following pedagogical-didactical "learning design" and "assessment" quality criteria in relation to OER?*	Matrix question, nine individual items, four-point importance scale
	How important is the quality criterion "license" to you in regard to OER?*	Four-point importance scale
	Which license conditions are important to you when using OER?	Matrix question, four individual items, four-point importance scale
	What licensing conditions are important to you when publishing OER?	Matrix question, four individual items, four-point importance scale
	Which quality criteria do you think are also important in relation to OER?	Free text answer
Sample description	What gender would you classify yourself as?	Drop-down selection
	To which age group would you assign yourself?	Drop-down selection
	What best describes your current job position?	Drop-down selection
	What describes your experience with OER best?	Drop-down selection

* According to Zawacki-Richter and Mayrberger [5].

as a link via mail. The survey period was from mid-April 2021 to mid-May 2021. In total, 157 people started the questionnaire, of which 89 respondents completed the questionnaire in full. Only the fully completed questionnaires were considered in the analysis of the data. Information on the sample can be seen in Table 3.

The sample is composed of 44 females and 40 male participants in total, 5 persons did not give any information about their gender. Table 3 shows that more than 50% of

Table 3. The sample

| | | Total: n=89 | |
		n	%
Gender			
	Female	44	49.44
	Male	40	44.94
	No specification	5	5.62
Age group			
	Up to 30 years	15	16.85
	31-40 years	28	31.46
	41-50 years	24	29.97
	51-60 years	16	17.98
	61-70 years	5	5.62
	Above 70 years	1	1.12
Profession			
	Research assistant	48	53.93
	Professor	17	19.10
	Post-doc	9	10.11
	Self-employed	4	4.49
	Private lecturer	3	3.37
	Employee in a commercial enterprise	2	2.25
	Retired / pensioned	1	1.12
	Other	5	5.62
OER experience			
	OER usage	5	5.62
	OER creation	29	32.58
	OER usage & creation	53	59.55
	No OER use or creation	2	2.25

the participants are in the age groups 31–40 years and 41–50 years. 22 respondents are older than 50 years and 15 participants are in the age group up to 30 years. Regarding the professional groups, the research revealed that 48 participants (53.93%) are working as research associates. This is followed by professors (17 participants), post-docs (9 participants) and others (5 participants). Also represented are self-employed persons, private lecturers, employees in companies and one retired person. The majority of respondents stated that they both use and create OER (n = 53). This was followed by creating OER with 29 participants and using OER with only 5 participants.The analysis was carried out using two analytical methods. The qualitative content analysis (QCA) according to Mayring [20] was used to analyze the open questions. Here, we divided the process of analysis into four phases. During the transcription process the researchers became familiar with the data, a rough word count and a classification was created to identify the criteria. In the first phase the answers were organized and then paraphrased (phase 1). Then, the paraphrases were generalized to a level of suitable abstraction into core sentences (phase 2). In the third phase the first reduction was made by cutting semantically identical core sentence and those which are not felt to add substantially to the content. Finally, as the second reduction, the core sentences were combined with similar or identical ones and thus classified in categories (phase 4). In naming the categories, we used the quality categories from the given model to ensure comparability. For the closed items, the mean values and standard deviations were analyzed using the statistical tool SPSS. The results are presented below.

4 Results

4.1 Quality of OER I – Open Questions

The questionnaire begins with two open, exploratory questions. The questions are assigned to question group "Quality of OER I" (see Table 2). The first question asks about the importance of quality in OER in general. In this way, the definition of quality of OER from the lecturer's perspective should be obtained. The second question is specifically directed at the quality requirements that lecturers personally place on their own teaching materials. The aim was to investigate whether the personal quality criteria differ from the general definition of the term. The results can be seen in Fig. 1.

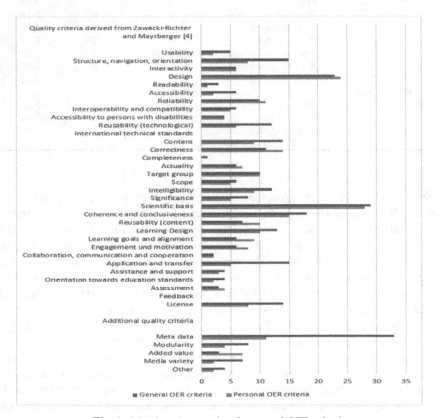

Fig. 1. Mentioned general and personal OER criteria

The presentation of the results of the open questions is divided into two parts. The first part shows the number of mentions for both questions, which were categorized according to the criteria overview of Zawacki-Richter and Mayrberger [5], while the second part shows a list with a number of additionally mentioned quality criteria, also from both questions. Respondents' mentions were counted only once so that duplicates were removed. If respondents' statements did not fit into one of the main or rather sub criteria, the sub category were used to classify them. This was the case for general statements on technical usability, technical reusability and technical accessibility, content quality and learning design. Another special aspect is that all mentions of the quality criterion target group were summarized under this term, since a differentiation as in Zawacki-Richter and Mayrberger (content-related target group and target group orientation) could not be made from the answers of the respondents.

For lecturers, meta data (n = 33) and the scientific basis (n = 29) are the upmost important quality criteria and thus significantly characterize the concept of quality from their point of view. Meta data means the overarching information about the particular resource to get a quick overview such as a description text or contact information from the author. In this way, lecturers see themselves supported in the search for suitable materials. Due to the frequency of mention, meta data was included as a separate criterion to highlight its importance. Therefore, we deviated from the Zawacki-Richter and Mayrberger model which sees meta data as a component under interoperability and compatibility [9]. Although metadata is mentioned in the definition of the quality criterion interoperability and compatibility, there is no emphasize on it. Since metadata was rated as very relevant by the respondents with 33 mentions in question 1 and 11 mentions in question 2, meta data should be a quality criterion on its own and should be clearly emphasized. The scientific basis of the resource is another prerequisite for the recognition of high-quality materials. These two criteria account for almost half of the mentions in the top six. Furthermore, the design (n = 28), the coherence and conclusiveness (n = 18), the application and transfer of knowledge (n = 15) as well as structure, navigation, orientation (n = 15) are important to the lecturers. By design, respondents mean the appealing visual presentation of OER, highlighted by professional and attractive layouts. The coherent content structure of the resource is described under coherence and conclusiveness. In addition, quality is determined by the applicability and the ability to transfer acquired knowledge from OER to new tasks, problems and contexts. Structure, navigation and orientation that influences the ease of use of the user interface is also an important quality criterion for the lecturers surveyed. The stimulation by the learning material for collaboration, communication and co-operation (n = 2) received the fewest mentions, as did completeness (n = 1). Whether OER are complete does not give any indication of quality for the respondents. Feed-back, i.e. the opportunity for learners to receive feedback on what they have learned, was not mentioned at all (n = 0). Complementary points to the quality criteria according to Zawacki-Richter and Mayrberger [5] are the already mentioned meta data as a separate category and modularity (n = 8). Through modularity, lecturers expect to be able to select and use content from the OER that is relevant to them without having to use the entire resource. This underlines the missing mentions of the completeness of the resource.

The first research question, what characterizes high-quality OER from the perspective of lecturers, can be answered with these results. From the perspective of lecturers, OER are considered to be of high quality if their metadata is complete, they are scientifically sound, coherent in their content and also user-friendly in their structure. It is noticeable that fewer statements were made in the second question concerning quality requirements. This can be attributed to the similarity of both questions, which may not have been perceived as distinct by some participants, so that they only briefly repeated themselves or only mentioned supplementary points. The results of the qualitative content analysis show large intersection of named categories. The personal requirements for OER also largely correspond to their understanding of high-quality OER. The two most frequently mentioned requirements are in the area of scientific basis (n = 28) and design (n = 24). Among the requirements for OER, the coherence and conclusiveness (n = 15) and the correctness (n = 14) of the content is also given high importance. Meta data (n = 11) is also found in the top 6, but was not mentioned as frequently as in the previous question. In addition, the topic of technical reliability (n = 11) is found in the most frequent mentions, which means robustness and technical stability in the context of the accessibility of the materials. It is striking that modularity (n = 4) is also reflected in the mentions of requirements. A large over-lap is also found in the missing mentions. In terms of requirements, completeness of content (n = 0), adherence to technical standards (n = 0), ability to provide feedback to learners (n = 0) and encouragement of collaboration, communication, and cooperation (n = 2) have little to no importance. These specifications are also reflected in the understanding of quality in OERs.

These results can be used to answer the second research question; which requirements are particularly relevant. The results indicate that lecturers want to use parts of OER and do not value complete teaching materials, rather they want OER to be modular. It is recommended to take out completeness as a criterion in the context of quality assurance and replace it with modularity instead. In addition, meta data should be given its own status by being included as a separate criterion in the quality assurance of OER. At the same time, the results reveal other items that are redundant from the perspective of the lectures and can be cut for quality measurement. These include adherence to technical standards, ability to provide feedback to learners, encouragement of collaboration, communication, and cooperation.

4.2 Quality of OER II – Closed Questions

In this section, the closed items of question group "Quality of OER II" (see Table 2), in which the participants were asked to rate the importance of the given quality criteria on a four-point Likert scale, are to be examined. The data were coded in such a way that 'not important' represents a 0, while 'very important' represents a 3. The highest possible value of importance is therefore a 3. With regard to the second research question, we set the value at which a quality criterion can be considered particularly relevant at 2.5. This value represents the midpoint between rather important (2) and very important (3) and thus represents clear agreement for high relevance from the perspective of the respondents. These mean values are displayed bold in Table 4. The standard deviation is also included in the interpretation of the data. A low standard deviation is to be interpreted here as an indication that the answers do not deviate far from the mean value and therefore the tendency of all respondents is towards the mean value. Here, a standard deviation of 0.5 and below is considered particularly meaningful. A SD up to 0.8 can considered sufficient for this study. A larger standard deviation could result in a larger standard error of the mean and a less precise estimate of the population mean. Table 4 provides an overview of the mean values (M) and standard deviations (SD) of the closed items.

Table 4. Means and standard deviations of the closed questions

Quality category	Quality criteria	M	SD
Technological criteria	Structure, navigation, orientation	**2.51**	0.71
	Interactivity	1.60	0.82
	Design	1.96	0.78
	Readability	**2.63**	0.55
	Reliability	**2.64**	0.57
	Interoperability and compatibility	2.17	0.76
	Accessibility to persons with disabilities	2.09	0.75
	International technical standards	1.83	0.83
Pedagogical-didactic criteria (content)	Correctness	**2.87**	**0.43**
	Completeness	1.76	0.75
	Actuality	2.18	0.65
	Target group	2.16	0.78
	Scope	1.81	0.74
	Intelligibility	**2.69**	**0.49**
	Significance	2.24	0.69
	Scientific Basis	**2.55**	0.60
	Coherence / conclusiveness	**2.55**	0.54
	Reusability	2.24	0.69
Pedagogical-didactic criteria (learning design & assessment)	Target group orientation	2.09	0.70
	Learning goals and alignment	2.20	0.79
	Engagement and motivation	2.35	0.66
	Collaboration, communication and cooperation	1.84	0.69
	Application and transfer	2.08	0.73
	Assistance and support	1.87	0.71
	Orientation towards education standards	1.60	0.84
	Assessment	1.52	0.84
	Feedback	1.75	0.79
Private International Law criteria	License	2.27	0.85

n=89, M=mean value, SD=standard deviation

As can be seen from the table above, criteria from the quality category technological criteria and pedagogical criteria (content) are classified as particularly important. With regard to the pedagogical criteria (learning design & assessment) and the licensing criterion, no quality criteria can be identified that were rated as particularly important. The highest mean value is achieved by the quality criterion correctness with 2.87, which also has the lowest standard deviation (SD = 0.43) of the whole survey. The second highest mean value is intelligibility with 2.69, which also has the second lowest standard deviation (SD = 0.49). The third and fourth places are held by reliability (M = 2.64, SD = 0.57) and readability (M = 2.63, SD = 0.55). Here, it should be noted that these are very close in terms of mean and standard deviation. In addition, the standard deviation for both of these quality criteria only just misses the threshold for being considered particularly meaningful. The fifth place is shared by coherence/conclusiveness (M = 2.55, SD = 0.54) and scientific basis (M = 2.55, SD = 0.60), although it should be noted that the standard deviation of the former is even lower. The last quality criterion, which still falls in the range of particularly important is structure, navigation, orientation with a mean value of 2.51 and a standard deviation of 0.71. Even if it did not reach the defined limit value, the following quality criterion engagement and motivation (M = 2.35, SD = 0.66) should be also mentioned here. The three quality criteria with the lowest mean values are orientation to educational standards (M = 1.60, SD = 0.84), interactivity (M = 1.60, SD = 0.82) and, last, evaluation (M = 1.52, SD = 0.84). According to the respondents, these three criteria play a rather subordinate role and were not rated as particularly relevant indicators of high quality in the context of OER.

5 Discussion

The basis for the model used by Zawacki-Richter and Mayrberger [5] are various established models for quality assurance of OER. However, little is known about the reliability and validity of the instruments [5]. The number of evaluation criteria used by the different models varies widely. In addition, some use hierarchically arranged quality dimensions and others use unsorted lists of criteria. The context of application ranges from generic approaches to specific subject areas. The model according to Zawacki-Richter and Mayrberger [5] provides a good cross-section of these models, as they have been compared and then the most important criteria have been elaborated. For this reason, this paper builds on this model. For this purpose, 89 lecturers in Germany at various institutions in the higher education sector were surveyed on the topic of quality measurement of OER, as the inclusion of lecturers has not yet been carried out empirically. In this context, the different roles of lecturers, as importers or exporters of OER, were further addressed. The respondents were identified by providing their own OER in repositories and were contacted via mail. Accordingly, the target group had experience with the creation and with the use of OER. The involvement of teachers is of great importance in this case, as they are the main producers and want to measure the quality of the materials. In addition, when implementing a quality assurance tool, it is important to design it in such a way that it meets the greatest possible acceptance among lecturers.

The survey shows that in the open questions design, a good scientific basis, as well as a high coherence and conclusiveness within the OER are very decisive for a high quality. The naming in both questions confirms the importance of these quality criteria. Furthermore, these criteria are also supported by the mean values of the statistical evaluation. However, the criterion of design was not given as much importance here, but rather a higher value was placed on technical reliability, readability, correctness of content and intelligibility. This deviation could be due to the fact that criteria such as readability, reliability, correctness or intelligibility are taken for granted as scientific standards. For this reason, these aspects may have been given less consideration in the free responses, but may have been ranked very high when the relevance was specifically queried. In general, the technical criteria were often perceived as important and thus frequently achieved a mean value above 2. In the open questions, another quality criterion was also frequently named, the meta data. The respondents often named a high transparency in the provision of contact information of the authors, as well as detailed description texts and information about the content of the OER. This information is contained in well-maintained meta data and would further increase the usability of OER according to the survey. Thus, a recommendation would be to classify and add the criterion meta data in the usability section. Zawacki-Richter and Mayrberger mention the term meta data in their definition of the quality criterion interoperability and compatibility, but metadata is not assumed in this criterion. Interoperability and compatibility were rarely mentioned, especially in the open questions, while meta data was explicitly assigned an important meaning through frequent naming. In contrast, the quality criteria completeness, feedback, assessment and orientation to educational standards were not mentioned at all or barely mentioned in both the open and closed questions. Here one could question the importance of the criteria and possibly reduce the model by excluding these criteria. With regard to the dual role of lecturers as importers or exporters, it was found that a large proportion of 60% of all respondents create and use OER. This makes a clear distinction of quality requirements based on their role difficult and cannot be clearly emphasized. On the other hand, the finding that only five respondents use OER but do not create it themselves and 29 people create OER but do not use other OER was surprising. Accordingly, there is obviously a greater willingness to share one's own materials than to implement other people's OER in one's own teaching. This finding supports the assumption that lecturers trust the quality of their own materials more.

Through the survey with lecturers, it was possible to show which quality criteria arc crucial for the quality assurance of OER. When creating OERs, lecturers should therefore pay particular attention to an appealing preparation of their materials and provide important basic information such as metadata, references to sources used, and a table of contents for a clear structure. A high degree of transparency and comprehensibility is crucial, so that other lecturers are more willing to embed external OERs in their own teaching. Nevertheless, the quality criteria are only one instrument for quality assurance and do not claim to cover the entire quality assurance process. A complete quality assurance process includes several phases and for some phases other indicators than quality criteria might be crucial. For this reason, it would make sense to continue this paper with further research and to develop a process model for the complete quality assurance process. Furthermore, the quality criteria considered here and the model according to Zawacki-Richter and Mayrberger do not provide a scoring model, so that an exact measurement of the criteria is not possible. For good general applicability, such a scoring model would be useful and represents another opportunity for further research.

References

1. Gulati, S.: Technology-enhanced learning in developing nations: a review. Int. Rev. Res. Open Dist. Learn. **9**(1), 346 (2008)
2. Ebner, M., Köpf, E., Muuß-Merholz, J., Schön, M., Schön, S., Weichert, N.: Ist-Analyse zu freien Bildungsmaterialien (OER) – Die Situation von freien Bildungsmaterialien in Deutschland in den Bildungsbereichen Schule, Hochschule, berufliche Bildung und Weiterbildung, Book on Demand, Norderstedt (2015)

3. UNESCO-Kommission: Pariser Erklärung zu OER, Weltkongress zu Open Educational Resources, UNESCO, Paris (2012). https://www.unesco.de/sites/default/files/2018-05/Par iser%20Erkl%C3%A4rung_DUK%20%C3%9Cbersetzung.pdf. Accessed 20 May 2021

4. Butcher, N., Malina, B., Neumann, J.: Was sind Open Educational Resources? und andere häufig gestellte Fragen zu OER. UNESCO, Bonn (2013)

5. Zawacki-Richter, O., Mayrberger, K.: Qualität von OER, Sonderband zum Fachmagazin Synergie (2017)

6. Yuan, L., MacNeill, S., Kraan, W.: Open Educational Resources – Opportunities and Challenges for Higher Education, JISC CETIS (2008)

7. OECD: Giving Knowledge for Free. The Emergence of Open Educational Resources, Centre for Educational Research and Innovation, pp. 30–31 (2007)

8. Butcher, N., Kanwar, A., Uvalić-Trumbić, S.: A basic guide to open educational resources (OER), vol. 2: Commonwealth of Learning; UNESCO, Section for Higher Education Verlag, Vancouver, Paris (2015)

9. Deimann, M., Neumann, J., Muuß-Merholz, J.: Open Educational Resources (OER) an Hochschulen in Deutschland – Bestandsaufnahme und Potenziale 2015, Whitepaper, open-educationale-resources.de (2015)

10. Ehlers, U.-D.: Extending the territory: from open educational resources to open educational practices. J. Open Flexible Dist. Learn. **15**(2), 1–10 (2011)

11. Brüggemann, H., Bremer, P.: Grundlagen Qualitätsmanagement. Von den Werkzeugen über Methoden zum TQM, Springer Vieweg, Wiesbaden (2020). https://doi.org/10.1007/978-3-658-28780-1

12. DIN EN ISO 9000: Quality management systems - Fundamentals and vocabulary (ISO 9000:2015), German and English version EN ISO 9000 (2015)

13. Lin, C., Wa, C.: Managing knowledge contributed by ISO 9001:2000. Int. J. Qual. Reliab. Manag. **22**, 968–985 (2005)

14. O´Mahony, K., Garavan, T.N.: Implementing a quality management framework in a higher education organization. A case study. Qual. Assurance Educ. **20**(2), 184–200 (2012)

15. Owlia, M.S., Aspinwall, E.M.: TQM in higher education – a review (1997)

16. Yuan, M., Recker, M.: Not all rubrics are equal: a review of rubrics for evaluating the quality of open educational resoruces. Int. Rev. Res. Open Distrib. Learn. **16**(5), 16–31 (2015)

17. Yu, X., Khazanchi, D.: Using embedded mixed methods in studying IS Phenomena: risks and practical remedies with an illustration. Commun. Assoc. Inf. Syst. **34**, 555–595 (2017)

18. Dhanapati, S.: Explanatory sequential mixed method design as the third research community of knowledge claim. Am. J. Educ. Res. **4**(7), 570 (2016)

19. Chyung, S.Y., Roberts, K., Swanson, I., et al.: Evidence-based survey design: the use of a midpoint on the Likert scale. Perform. Improv. **56**(10), 15–23 (2017)

20. Mayring, P.: Qualitative content analysis: theoretical foundation, basic procedures and software solution. SSOAR, Klagenfurt (2014)

Using Wearable Fitness Trackers to Detect COVID-19?!

Christina Gross[✉] ⓘ, Wladimir Wenner ⓘ, and Richard Lackes ⓘ

TU Dortmund, Otto-Hahn-Str. 12, 44227 Dortmund, Germany
{christina.gross,wladimir.wenner,richard.lackes}@tu-dortmund.de

Abstract. Wearable Fitness Trackers (WFTs) could both lead to healthier lifestyles and be suitable for tracking infectious diseases. However, though they have existed for many years they are still not prevalent. We developed a research model which was theoretically embedded in technology acceptance research, then analyzed the data collected through a survey of 207 WFT users and non-users. Our findings suggest that, in respect to both users and non-users, perceived value is the strongest driver of usage intention. Furthermore, we found significant differences between male and female participants. Unfortunately, we were not able to confirm a relation between Covid-19 awareness and WFT usage intention. Our study provides relevant findings to manufacturers and insights into the consumer WFT needs.

Keywords: Wearable Fitness Trackers · Covid-19 · Technology acceptance

1 Introduction

Modern wearable technology has been influenced by the evolution of ubiquitous computing and the Internet of Things. One of the first products used for activity tracking was Fitbit, which was released 2009 [1]. In the following years, smartwatches were released by all the major electronic companies, such as Samsung's Galaxy Gear in September 2013 [2] and the Apple Watch in April 2015 [3]. Wearable Fitness Trackers (WFT) track or monitor the physical activity of the person wearing them [4]. Although they have been around for many years, they are not widely used. In 2019, only 29% of Germans owned such a device [5]. Our first research question is: Which factors foster intention to use wearable fitness trackers? (**RQ1**).

Worldwide, 2020 was dominated by the COVID-19 pandemic, which was caused by severe acute respiratory syndrome Coronavirus 2 (SARS-CoV-2). WFTs and their sensor data can record changes in heart rate, sleep duration and physical activity. In this way, they are suitable for supporting counts of the rate of new infections and thus hindering the dissemination of the disease. The Robert Koch Institute is both monitoring and estimating risks for the population in Germany and doing COVID-19 research. For this purpose, people can donate their WFT data and contribute to research [6]. In 2020, the German press reported on this possibility. However, it is questionable whether this trend might lead to a higher utilization of WFT. Consequently, our second research question will be: To which extent is the intention to use wearable fitness trackers influenced by COVID-19 awareness? (**RQ2**).

© Springer Nature Switzerland AG 2021
R. A. Buchmann et al. (Eds.): BIR 2021, LNBIP 430, pp. 51–65, 2021.
https://doi.org/10.1007/978-3-030-87205-2_4

2 Background and Related Literature

WFTs are suitable for stimulating a user's physical activity, as most contain a pedometer and record athletic training. In this way, they are suitable for improving training and starting to exercise. The usage intention and reasons for using or not using a certain technology have been examined with manifold models (Table 1).

Table 1. Literature research

Authors	Focus area	Findings
Baskaran and Mathew 2020 [7]	Wearable users' privacy coping	Perceived privacy risk influences efficacy of WFT
Lunney et al. 2016 [8]	Acceptance and perceived outcomes of WFT	Perceived usefulness and perceived ease of use are key drivers of acceptance. A higher use of WFT leads to a healthier lifestyle
Guillén-Gámez and Mayorga-Fernández 2019 [9]	Perceptions of patients and relatives about the acceptance of wearable devices to improve their health	Interest in WFT is higher for males, but is rising for females. Privacy protection is strongly influencing usage
Yang et al. 2016 [10]	User acceptance of wearable devices: an extended perspective of perceived value	Using WFT depends strongly on perceived value. Users value perceived (financial and performance) risk not as important, as non-users of WFT
Chuah et al. 2016 [11]	Role of usefulness and visibility in smartwatch adoption	Perceived usefulness and visibility mainly influence adoption, depending on if WFT are either regarded as accessories or health improving technology
Pfeiffer et al. 2016 [12]	Consumer acceptance of wearable self-tracking devices	Trust (in provider) plays an important role. It depends on the user's age if a device is regarded as toy or health supporting
Asadi et al. 2019 [13]	Determinants of adoption of wearable healthcare devices	Trust in devices and interest in own health are most important for consumers. Users trust their devices

The nationwide lockdown because of Covid-19 has led to less physical activity [14], which can be monitored by WFTs. Users might increase their immunity with an increase in motion and physical activity. Because gyms have been closed, a lot of people have bought fitness devices for the home. It is likely that their WFT usage has increased. The ability to detect a potential infection with domestic possibilities would also solve the governmental problem of limited test capacities. Furthermore, the spreading of the virus could be hindered if people can recognize an early shift in their vital parameters and therefore reduce their contact to other people. This paper is one of the first to empirically explore the relationship between Covid-19 awareness and WFTs.

3   Theoretical Perspective and Hypothesis Development

Most empirical studies predicting the acceptance of a technology are based on the Technology Acceptance Model (TAM) [15]: In this model, Perceived Usefulness (PU) and Perceived Ease of Use (PEOU) influence the Attitude towards Using (A). This Attitude influences the Behavioral Intention (BI) to use and thus the Actual Usage (AU) of a certain technology. The model could be extended by external factors which influence PU and PEOU [15]. Current WFT research extended and applied TAM [8, 11, 16]. Since WFTs have been around for many years, we propose a model based on TAM, but extended and modified with elements from current research.

3.1   Covid-19 Awareness (CA) and Behavioral Intention (BI)

CA is interpreted in many ways in existing research. There are studies examining the perception of the pandemic [17], while others focus on virus knowledge [18]. This study stresses the perception of the COVID-19 pandemic. If people worry about infections, they might use WFTs. Using WFTs could provide information about such an infection [19]. It is thus more relevant to elderly people, because when infected, their risk is higher. BI describes the subjective probability that a person behaves in a certain way [20]. Unlike TAM, we will use this construct as our target construct, because we will focus on intention more than on attitude and actual use of WFTs because of the low assumed user rate. Therefore, we hypothesize:

Covid-19 Awareness positively influences the intention to use WFTs (**H1**).

3.2   Perceived Usefulness (PU)

PU is the degree to which a person thinks that using a specific system would exaggerate their performance [15]. The positive relation between PU and BI for WFTs has often been proven [8, 21, 22]. We will integrate this construct as an intersection between TR and BI:

Perceived Usefulness positively influences the intention to use WFTs (**H2**).

3.3 Perceived Risk (PR)

Consumer behavior cannot be foreseen, such as how every action bears consequences [23]. This leads, according to the Theory of Planned Behavior (TPB), to a lower likelihood of a person performing a certain action [24]. In our context, someone would use WFTs if they assume a low risk and high behavior control. Risks can be seen in financial, social or psychological ways. Some authors focused on several risk aspects [25], but we will focus on privacy risks and hypothesize:

Perceived Risk negatively influences the intention to use WFTs (**H3a**).

Moreover, a relationship between PR and PV is likely. If a consumer assumes a low risk will come from using a particular technology, the value of this technology might be perceived as higher. This relationship is proven for cell phones [26] and we will propose it for WFTs:

Perceived Risk negatively influences the perceived value of WFTs (**H3b**).

3.4 Perceived Value (PV)

There are four definitions of PV [27]: A low price, a desire or expectation of the consumer, the quality-price relationship, and the sum of these factors. We will define PV as the expectation of WFT users, according to Yang et al. (2016) [10]. Unlike in their research, we will vary the constructs preceding PV:

Perceived Value positively influences the intention to use WFTs (**H4**).

3.5 Trust (TR)

Trust is described as a relationship between different parties, in which each party is interested in the wellbeing of the other and considers the consequences for the other party when making decisions [28]. This research focuses on consumer trust in manufacturers. When questioning our participants, we also asked if they trusted their specific WFT. In literature, the relationship between TR and WFTs was brought up often. In this context, a study divided trust into several categories (Trusting Beliefs, General Trust, Trusting Intentions, Manufacturer Trusting Perceptions) and found correlations between these constructs [29]. This author performed research on WFT and trust, but did not focus on the relationship between TR, PU, PR and PV. Other authors examined the relationship between TR and PU for technology acceptance in the public health sector [30]. In our study, we will focus on WFTs, and our hypothesis is:

Trust in WFT manufacturers positively influences the Perceived Usefulness (**H5a**).

Moreover, it is obvious that TR and Perceived Risk (PR) are related. Some authors found the risk to be lower when trust in the provider is higher [31] (in this case, the seller). This could be transferred to our case, in which we will focus on privacy risks:

Trust in technology and manufacturers negatively influences the Perceived Risk (**H5b**).

We will also focus on the Perceived Value (PV) of a WFT. It is likely that high value and great trust in the manufacturer are related, which has already been proven for the E-Commerce field of application [32]. Consequently, we hypothesize:

Trust in technology and manufacturers positively influences Perceived Value of WFT (**H5c**).

3.6 Subjective Norm (SN)

The SN is derived from the theory of reasoned action [20]. SN describes someone's perception of a specific issue. It focuses therefore the person and their social environment. This concerns how the individual's behavior is dependent on expectations which exist in their social environment and how the individual complies with this expectations [20]. WFT use is positively related to SN [8]. Furthermore, according to our literature research, we will examine the relationship between SN and TR. In another context, a positive relationship has been discovered between SN and TR. This is why our hypothesis is:

The Subjective Norm positively influences trust in WFT manufacturers (**H6a**).

3.7 Perceived Enjoyment (PE)

PE is derived from TAM [15], and represents intrinsic motivation and thus the intention (BI) to use a certain technology. Existing WFT research has confirmed the positive relationship between PE and BI [12]. When comparing the age of the participants, this effect was not significantly proven for people under 25 [12]. We will examine the effect of PE on TR, which was examined for mobile payment before [33]. We assume that consumers have a higher trust in manufacturers if they enjoy using their product. This might indicate that their interests are in the manufacturer's focus, and we thus hypothesize:

Perceived WFT Enjoyment positively influences Trust in WFT manufacturers (H6b).

3.8 Proposed Research Model

The behavioral intention to use WFT should be determined by Covid-19 Awareness (H2) and PU, PR and PV and their antecedents TR, SN and PE (H1). Figure 1 shows our research model.

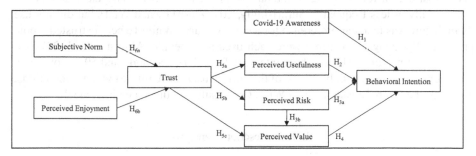

Fig. 1. Research model

4 Empirical Methods

First, we conducted a literature review to obtain preliminary variables for factors affecting WFT usage. We then specified the structural model and developed a questionnaire to measure the required data. Our questionnaire was generated with LimeSurvey and distributed on social media and SurveyCircle. The sampling technique used was non-probability sampling. After data collection, the Partial Least Squares (PLS) method was applied and group differences were examined. The data fulfilled the sample size recommendations for stable results of the model estimation [34]. Finally, we interpreted our results.

The questionnaire, distributed between October and November 2020, contained 38 questions for the model and nine demographic and general questions. Our measurement items were captured by a 5-point-Likert scale (1 = completely disagree; 5 = fully agree). All the measurement instruments were adapted from published research. The constructs and their corresponding items are displayed in Table 2. In total, 207 questionnaires were submitted. The empirical data was analyzed with Smart PLS 3.3.2 [35], which is suitable for smaller sets and is a non-parametric method. Bootstrapping was done with 5,000 samples to determine the significance of weights, loadings and path coefficients [36]. For missing values, case-wise replacement was applied. The program was set to 300 maximum iterations, with a stop criterion of 7.

4.1 Statistical Analyses and Results

Descriptive Analysis. Our sample contained an almost equal number of male and female German participants (94 and 113). The majority of the participants were under 30 (70%) and students (63.3%). 69 (33.3%) of our participants were in employment. The fact that our participants were either highly qualified (48.3%) or students led to a high education level in our sample. Only 23.7% (49) of the participants had a net income below €10,000, which led to the assumption that they either still lived with their parents or with other students in a single household. 13.5% (28) stated that their net income was higher than €70,000.

Most of the participants knew about WFTs, but just 37.2% (77) used and owned one. Most used it for more than 2 h a day (42.9%), 11.7% (9) almost daily and 24.5% (19) weekly or less frequently. Surprisingly, 20.8% (16) owned WFTs but did not use them. There was no clear preference for a certain brand. While 16 people trusted Apple products, Samsung and Garmin were each used by 9 people. Almost a third (24) used another manufacturer. The purchase intention for WFTs was high, and 59% would buy (another) WFT. Interestingly, 23.4% of the current users have future purchase intentions, although they were already using a WFT. Our results are displayed in Table 2.

Table 2. Descriptive analysis

Dimension and level	Frequency	Percentage Total	Percentage Owner	Non-owner	Dimension	Frequency	Percentage Total
Gender					Ownership		
Male	94	45.4	42.9	46.9	Yes	77	37.2
Female	113	54.6	57.1	53.1	No	130	62.8

(continued)

Table 2. (*continued*)

Dimension and level	Frequency	Percentage Total	Percentage Owner	Non-owner	Dimension	Frequency	Percentage Total
Age					Usage time		
19–24	94	45.4	35.0	51.5	>2 h per day	33	42.9
25–29	51	24.6	26.0	23.9	<2 h per day	9	11.7
30–39	38	18.4	24.7	14.6	At least weekly	14	18.1
40–49	13	6.3	7.8	5.4	More rarely	5	6.5
>49	11	5.3	6.5	4.6	No usage	16	20.8
Academic qualification					Brand		
GCSE	13	6.3	9.1	4.6	Apple	16	20.8
A levels	94	45.4	40.2	48.5	Fitbit	13	16.9
Bachelor	67	32.3	27.3	35.4	Huawei	6	7.8
Master	31	15.0	20.8	11.5	Samsung	9	11.7
Doctorate	2	1.0	2.6	0	Garmin	9	11.7
Occupation					Others	24	31.1
Student	131	63.3	49.3	71.6			
Professional	61	29.5	40.3	23.1			
Job-seeker	6	2.9	5.2	1.5			
Self-employed	3	1.4	1.3	1.5			
Marginal Employment	6	2.9	3.9	2.3			
Net income							
<10.000 EUR	49	23.7	16.9	27.6			
10.000–29.999 EUR	51	24.6	20.7	27			
30.000–49.999 EUR	44	21.3	26.0	18.5			
50.000–69.999 EUR	35	16.9	18.2	16.1			
>70.000 EUR	28	13.5	18.2	10.8			
Purchase intention							
Yes	59	28.5	23.4	31.5			
No	148	71.5	76.6	68.5			

The structural equation model (SEM) consists of two elements: the measurement model, which specifies the relationship between the constructs and their indicators, and the structural model, in which the relationships between the constructs are analyzed [34, 37]. The PLS-SEM approach was chosen over a covariance based analysis because our research question is exploratory, and because of its proven power in explaining complex behavior research [38].

Assessing the Measurement Model. Within the measurement model, we distinguished between reflective and formative constructs. The construct Covid-19 Awareness is measured formatively, while the other constructs were measured reflectively [39].

Our results and the measurement instruments are displayed in Table 3.

Table 3. Operationalization of constructs

Construct	Items	Questionnaire item focus	Mean	SD	Weights	VIF	Source
Perceived Enjoyment (PE)	PE1	Fun	3.26	1.230	0.260***	6.490	[40]
	PE2	Pleasantness	3.33	1.190	0.274***	6.817	[41]
	PE3	Excitement	3.66	1.080	0.234***	2.651	[42]
	PE4	Entertainment	3.24	1.056	0.136***	2.322	
	PE5	Happiness	2.80	1.027	0.259***	2.152	[43]
Subjective Norm (SN)	People …						
	SN1	Influencing my behavior	2.14	1.098	0.212**	5.286	[44]
	SN2	Important to me	2.10	1.084	0.217***	7.936	
	SN3	In my social circle	2.09	1.096	0.193**	8.085	
	SN4	Using WFT would enhance my status	1.79	0.908	0.299*	1.795	[45]
Trust (TR)	WFT manufacturers …						
	TR1	Are honest	3.15	0.771	0.337***	1.528	[44]
	TR2	Collect reliable data	3.47	0.891	0.310***	1.378	
	TR3	Keep customers best interests in mind	3.08	0.962	0.344***	1.286	[33]
	TR4	Keep their promises	3.38	0.719	0.337***	1.606	[30]
Perceived Risk (PR)	Using a WFT …						
	PR1	Is risky (health data)	2.90	1.117	0.267***	1.870	[44]
	PR2	Infringes on my privacy	2.67	1.131	0.218***	2.622	[45]
	PR3	Feels secure	3.39	0.938	0.419***	1.373	
	PR4	Have more uncertainties cf. other technologies	2.72	0.922	0.214***	1.523	[46]
	PR5	Leads to loss of control (personal data)	2.67	1.102	0.191***	2.018	[25]

(continued)

Table 3. (*continued*)

Construct	Items	Questionnaire item focus	Mean	SD	Weights	VIF	Source
Perceived Value (PV)	Using a WFT …						
	PV1	Is beneficial	4.00	0.700	0.228***	1.347	[47]
	PV2	Leads to disadvantages	2.91	1.060	0.202***	1.101	
	PV3	Meets my expectations	3.63	0.802	0.259***	1.518	[48]
	PV4	Offers great value	3.11	1.042	0.324***	2.075	
	PV5	Represents good use of time and money	3.12	1.117	0.352***	2.354	[49]
Perceived Usefulness (PU)	A WFT …						
	PU1	Is overall useful	3.71	0.997	0.322***	2.029	[31]
	PU2	Will make it easier to take care of health	3.87	0.944	0.273***	3.095	[30]
	PU3	Helps staying healthy	3.73	1.015	0.265***	3.121	Own research
	PU4	Saves time	2.52	0.929	0.089***	1.138	[50]
	PU5	Improves my sports performance	3.77	0.992	0.257***	2.180	[51]
Behavioral Intention (BI)	Using a WFT…						
	BI1	Is worthwhile	3.52	1.014	0.255***	2.346	[40]
	BI2	Frequently	2.92	1.224	0.208***	2.397	
	BI3	Likely in future	3.12	1.364	0.232***	3.611	[42]
	BI4	Is an option to me	3.35	1.252	0.222***	2.780	
	BI5	Is recommendable	3.12	1.182	0.232***	3.140	[52]
Covid Awareness (CA)	COVID-19						
	CA1	An infection is likely	3.02	1.209	0.105ns	1.353	[17]
	CA2	I think, I would get sick, if infected	3.03	1.038	0.060ns	1.161	
	CA3	Likely infection of family and/or friends	3.66	1.063	0.188ns	1.134	
	CA4	Preparedness to fight it	4.23	0.921	0.559***	1.295	[53]
	CA5	Affected from strong restrictions in daily life	3.73	1.071	0.551***	1.079	[18]

Significance of indicators: ns = not significant; *p < 0.10, **p < 0.05, ***p < 0.01.

The reflective measurement model was evaluated using four criteria: the internal consistency reliability (ICR), the indicator reliability (IR), the convergent validity (CV), and the discriminant validity (DV). For the ICR, Cronbach's Alpha (CA) exceeds 0.70 and the composite reliability (CR) exceeds 0.80, indicating solid reliability [54]. For the

CR, the Average Variance Extracted (AVE) values exceed 0.70 [55]. For the DV, the Fornell-Larcker criterion (FLC) and the cross-loadings need to be checked. The FLC states that the AVE of each latent variable should be greater than the latent variable's highest squared correlation with any other latent variable [55]. Furthermore, the cross-loadings ensure that the models' constructs differ sufficiently from each other if each indicator loads the highest with its own items [56].

The formative measurement model was evaluated using three criteria: the convergent validity, the collinearity among indicators, and the significance and relevance of outer weights. Our redundancy analysis confirmed the convergent validity. The Variance Inflation Factor (VIF) is used as an indicator of collinearity. As shown in Table 3, most of our VIFs did not exceed the maximum level of 10 [57] and the threshold of 5 [58]. Finally, our outer weights were significant after bootstrapping.

Assessing the Structural Model. We first assessed the structural model for collinearity issues, as because there are some high VIF values, collinearity issues might exist. Then we assessed the significance and relevance of the structural model relationships. As shown in Fig. 2, all the path coefficients except SN → TR and CA → BI were significant. Therefore, H1 and H6a needed to be rejected, while all other hypotheses were confirmed. When assessing the endogenous constructs, we found three of five latent variables to have a low R^2 value. The explanatory power of TR, PU and PR was thus low, based on their preceding constructs. However, BI was explained with 73.8% based on its preceding constructs. We next assessed effect sizes f^2. The effect between PV and BI was high ($f^2 = 0.601$). Finally, the predictive relevance Q^2 showed a strong effect on our target construct BI, as shown in Fig. 2.

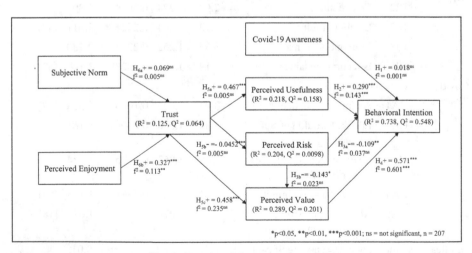

Fig. 2. Results of the research model

Multi-group Analysis (MGA). With the MGA we estimated our path model for male and female participants, discovering group-specific differences. We first tested for measurement invariance and eliminated indicator SN4 [59], which was possible because of reflective measurement. As an MGA result, the path (SN → TR) differed significantly (p = 0.027) for males and females. This led to a lower path coefficient for men (−0.169) than for women (0.163). This result was surprising, because H6a needed to be rejected. No significant group-specific differences were detected.

5 Discussion

5.1 Discussion of Findings

We found PU, PR, PV and PE to be fostering factors for WFT usage intentions (RQ1). In contrast to exiting literature, we refrained from using the traditional attitude construct and measured the direct relation between usefulness and usage intention [8, 11]. Adding attitude as a construct, it will have a high relation to BI. Moreover, the PR plays a minor role [9]. One reason for this might be that we focused on privacy concerns while they measured PR with just two indicators, one of them measuring the reliability of the device. Besides this, the usage intention is higher when the PR is lower. Essentially, the PV of WFTs has the highest impact on usage intention, which is comparable to Yang et al. (2016) [10]. Therefore, consumers need beneficial functions. This seems to be more important to them than higher usefulness or a lower risk. This trend can be explained by the fact that WFTs have existed for over 10 years and manufacturers have developed them as far as they possibly can. Moreover, the lower the PR is, the higher the PV will be. In our case, this effect was not strong. This implies that consumers do not take privacy concerns into account when thinking about WFTs. In the European Union, there is a General Data Protection Regulation (GDPR) to protect personal data. All of the three hypotheses which deal with trust in manufacturers were proven, albeit moderately. For the preceding constructs SN and PE, our results were ambiguous. We were thus not able to confirm the influence of an individual environment on trust. However, trust in manufacturers is positively influenced by individual's enjoyment. For our group comparison, we found that female participants took SN into account more than the male participants.

This paper is one of the first to examine the relationship between Covid-19 Awareness and WFTs (RQ2). It is likely that people who take care of their health intend to use WFTs. Also German media has reported on the opportunity to discover an infection with the help of an WFT. In addition to this, it would deliver insights to the Robert Koch Institute, but most people in Germany do not use this technology. Unfortunately, our hypothesis was not proven. In our study, we did not nudge our participants. If we had shown them an article covering this in advance, their answers might have been different.

5.2 Theoretical Contributions

Our proposed research model was proven, and eight out of the ten hypotheses were confirmed. Contributing to the research, we investigated the influence of the Covid-19 Awareness on WFT usage intention. This hypothesis could not be confirmed. Nevertheless, this result might provide a first step and lead to further consideration of the topic. Beyond this, the difference between male and female participants needs additional exploratory research for WFT and possibly the Internet of Things in general.

5.3 Implications for Practice

WFT manufacturers should emphasize the additional benefits of their products. For consumers, the added value of these devices is important. Consumers are not generally worried about privacy. It can be concluded that they do not see the advantage of these devices. Today, many smartphones offer the same functions as WFTs. This is why manufacturers should question the functions which are unique when using a separate WFT. Such a function might be that they are waterproof, and therefore suitable for recording swimming. In addition, they are lightweight and need less space than a smartphone. The possibility of detecting infection diseases by monitoring personal health data [19] might be used as a unique selling position. A lot of manufacturers already collect user data. They could donate this information to research, which would help in gaining new insights, and result in higher demand. Policy makers could support manufacturers because the data and insights gained could help in ending the Covid-19 pandemic. This might be a quick win, but it would help strengthen attention for the important topic of personal health. For consumers, using a WFT has a lot of advantages. The ability to track and monitor physical activity could help someone become more active. Manufacturers could support user needs with gamification elements, such as leaderboards and rewards for good behavior, such as points and badges. This could be supported with challenging but achievable tasks, which would foster personal training and lead to a healthier lifestyle.

6 Limitations and Future Research

As with any research project, this study has several limitations. Most of our participants were under 30 years old and had a high level of education. This is why our research needs further attention. It is possible that people with a lower level of education or limited financial capabilities might be less informed about WFTs, or focus on their costs. WFTs are suitable for any age group, though past research has identified that the elderly might have to overcome some obstacles in technology acceptance. This is why we propose a larger sample size to check on age-specific group differences. Moreover, our study language was German and therefore limited. There are nations that are less invested in technology, and it would be interesting to compare results with such countries. As proposed, the relation between WFT usage and Covid-19 might be stressed more, so that the buyers realize WFTs can be tools for a personal fight against it. Our construct SN needs more research. Our research would be more generalized if Covid-19 Awareness was substituted with health awareness.

References

1. The Consumer Electronics Hall of Fame: Fitbit - IEEE Spectrum. https://spectrum.ieee.org/consumer-electronics/gadgets/the-consumer-electronics-hall-of-fame-fitbit. Accessed 26 July 2021
2. Samsung Galaxy Gear auf der IFA 2013 offiziell vorgestellt. https://www.giga.de/zubehoer/samsung-smartwatch/videos/samsung-galaxy-gear-auf-der-ifa-2013-offiziell-vorgestellt/. Accessed 26 July 2021

3. Apple Watch Arrives in Seven More Countries June 26. https://www.apple.com/de/newsroom/2015/06/04Apple-Watch-Arrives-in-Seven-More-Countries-June-26/. Accessed 26 July 2021
4. Kao, Y.-S., Nawata, K., Huang, C.-Y.: An exploration and confirmation of the factors influencing adoption of IoT-based wearable fitness trackers. Int. J. Environ. Res. Public Health **16**, 3227 (2019)
5. Bitkom: Anteil der Fitnesstracker-Nutzer in Deutschland bis 2019. https://de.statista.com/statistik/daten/studie/1047599/umfrage/anteil-der-fitnesstracker-nutzer-in-deutschland/. Accessed 12 Feb 2021
6. Robert Koch Institute: Corona-Datenspende. https://corona-datenspende.de/. Accessed 12 Feb 2021
7. Baskaran, K., Mathew, S.K.: Danger vs fear: an empirical study on wearable users' privacy coping. In: Proceedings of the 2020 on Computers and People Research Conference, pp. 123–132 (2020)
8. Lunney, A., Cunningham, N.R., Eastin, M.S.: Wearable fitness technology: a structural investigation into acceptance and perceived fitness outcomes. Comput. Hum. Behav. **65**, 114–120 (2016)
9. Guillén-Gámez, F.D., Mayorga-Fernández, M.J.: Empirical study based on the perceptions of patients and relatives about the acceptance of wearable devices to improve their health and prevent possible diseases. Mob. Inf. Syst. **2019** (2019)
10. Yang, H., Yu, J., Zo, H., Choi, M.: User acceptance of wearable devices: an extended perspective of perceived value. Telemat. Inform. **33**, 256–269 (2016)
11. Chuah, S.H.-W., Rauschnabel, P.A., Krey, N., Nguyen, B., Ramayah, T., Lade, S.: Wearable technologics: the role of usefulness and visibility in smartwatch adoption. Comput. Hum. Behav. **65**, 276–284 (2016)
12. Pfeiffer, J., von Entress-Fuersteneck, M., Urbach, N., Buchwald, A.: Quantify-me: consumer acceptance of wearable self-tracking devices (2016)
13. Asadi, S., Abdullah, R., Safaei, M., Nazir, S.: An integrated SEM-Neural Network approach for predicting determinants of adoption of wearable healthcare devices. Mob. Inf. Syst. **2019** (2019)
14. Pal, R., Yadav, U., Verma, A., Bhadada, S.K.: Awareness regarding COVID-19 and problems being faced by young adults with type 1 diabetes mellitus amid nationwide lockdown in India: a qualitative interview study. Prim. Care Diabetes **15**, 10–15 (2021)
15. Davis, F.D.: Perceived usefulness, perceived ease of use, and user acceptance of information technology. MIS Q. **13**, 319–340 (1989)
16. Kim, K.J., Shin, D.-H.: An acceptance model for smart watches. Internet Res. (2015)
17. Nindrea, R.D., et al.: Survey data of COVID-19 awareness, knowledge, preparedness and related behaviors among breast cancer patients in Indonesia. Data Br. (2020)
18. Mandal, I., Pal, S.: COVID-19 pandemic persuaded lockdown effects on environment over stone quarrying and crushing areas. Sci. Total Environ. **732**, 139281 (2020)
19. Albert, H.: Wearable fitness trackers could help detect Covid-19 cases. Forbes (2020)
20. Fishbein, M., Ajzen, I.: Belief, attitude, intention, and behavior: An introduction to theory and research. Philos. Rhetoric **10**(2), 130–132 (1977)
21. Cheung, M.L., et al.: Examining consumers' adoption of wearable healthcare technology: the role of health attributes. Int. J. Environ. Res. Public Health **16**, 2257 (2019)
22. Koo, S.H.: Consumer differences in the United States and India on wearable trackers. Fam. Consum. Sci. Res. J. **46**, 40–56 (2017)
23. Bauer, R.A.: Consumer behavior as risk taking. In: Proceedings of the 43rd National Conference of the American Marketing Assocation, Chicago, Illinois, 15–17 June 1960. American Marketing Association (1960)
24. Ajzen, I.: The theory of planned behavior. Organ. Behav. Hum. Decis. Process. **50**, 179–211 (1991)

25. Featherman, M.S., Pavlou, P.A.: Predicting e-services adoption: a perceived risk facets perspective. Int. J. Hum. Comput. Stud. **59**, 451–474 (2003)
26. Snoj, B., Korda, A.P., Mumel, D.: The relationships among perceived quality, perceived risk and perceived product value. J. Prod. Brand Manag. (2004)
27. Zeithaml, V.A.: Consumer perceptions of price, quality, and value: a means-end model and synthesis of evidence. J. Mark. **52**, 2–22 (1988)
28. Kumar, N.: The power of trust in manufacturer-retailer relationships. Harv. Bus. Rev. **74**, 92 (1996)
29. Bui, M.: Trust in Internet of Things technology: wearable fitness devices (2020)
30. Dhaggara, D., Goswami, M., Kumar, G.: Impact of trust and privacy concerns on technology acceptance in healthcare: an Indian perspective. Int. J. Med. Inform. **141**, 104164 (2020)
31. Pavlou, P.: Consumer intentions to adopt electronic commerce-incorporating trust and risk in the technology acceptance model. In: Digit 2001 Proceedings, p. 2 (2001)
32. Ponte, E.B., Carvajal-Trujillo, E., Escobar-Rodríguez, T.: Influence of trust and perceived value on the intention to purchase travel online: integrating the effects of assurance on trust antecedents. Tour. Manag. **47**, 286–302 (2015)
33. Rouibah, K., Lowry, P.B., Hwang, Y.: The effects of perceived enjoyment and perceived risks on trust formation and intentions to use online payment systems: new perspectives from an Arab country. Electron. Commer. Res. Appl. **19**, 33–43 (2016)
34. Chin, W., Newsted, P.: Structural equation modeling analysis with small samples using partial least square. Stat. Strateg. Small Sample Res. (1999)
35. Ringle, C.M., Wende, S., Becker, J.-M.: SmartPLS 3. Boenningstedt SmartPLS GmbH (2015)
36. Hair, J.F., Jr., Hult, G.T.M., Ringle, C., Sarstedt, M.: A Primer on Partial Least Squares Structural Equation Modeling (PLS-SEM). Sage Publications, Thousand Oaks (2016)
37. Reinartz, W., Haenlein, M., Henseler, J.: An empirical comparison of the efficacy of covariance-based and variance-based SEM. Int. J. Res. Mark. **26**, 332–344 (2009)
38. Lowry, P.B., Gaskin, J.: Partial least squares (PLS) structural equation modeling (SEM) for building and testing behavioral causal theory: when to choose it and how to use it. IEEE Trans. Prof. Commun. **57**, 123–146 (2014)
39. Diamantopoulos, A., Winklhofer, H.M.: Index construction with formative indicators: an alternative to scale development. J. Mark. Res. **38**, 269–277 (2001)
40. Hsu, C.-L., Lin, J.C.-C.: Acceptance of blog usage: the roles of technology acceptance, social influence and knowledge sharing motivation. Inf. Manag. **45**, 65–74 (2008)
41. Mun, Y.Y., Hwang, Y.: Predicting the use of web-based information systems: self-efficacy, enjoyment, learning goal orientation, and the technology acceptance model. Int. J. Hum. Comput. Stud. **59**, 431–449 (2003)
42. Lew, S., Tan, G.W.-H., Loh, X.-M., Hew, J.-J., Ooi, K.-B.: The disruptive mobile wallet in the hospitality industry: an extended mobile technology acceptance model. Technol. Soc. **63**, 101430 (2020)
43. Lisha, C., Goh, C.F., Yifan, S., Rasli, A.: Integrating guanxi into technology acceptance: an empirical investigation of WeChat. Telemat. Inform. **34**, 1125–1142 (2017)
44. Gong, Z., Han, Z., Li, X., Yu, C., Reinhardt, J.D.: Factors influencing the adoption of online health consultation services: the role of subjective norm, trust, perceived benefit and offline habit. Front. Public Health **7**, 286 (2019)
45. Kaushik, A.K., Agrawal, A.K., Rahman, Z.: Tourist behaviour towards self-service hotel technology adoption: trust and subjective norm as key antecedents. Tour. Manag. Perspect. **16**, 278–289 (2015)
46. Im, I., Kim, Y., Han, H.-J.: The effects of perceived risk and technology type on users' acceptance of technologies. Inf. Manag. **45**, 1–9 (2008)

47. Kim, T.G., Lee, J.H., Law, R.: An empirical examination of the acceptance behaviour of hotel front office systems: an extended technology acceptance model. Tour. Manag. **29**, 500–513 (2008)
48. Chen, S.-Y., Lu, C.-C.: Exploring the relationships of green perceived value, the diffusion of innovations, and the technology acceptance model of green transportation. Transp. J. **55**, 51–77 (2016)
49. Kim, Y.H., Kim, D.J., Wachter, K.: A study of mobile user engagement (MoEN): engagement motivations, perceived value, satisfaction, and continued engagement intention. Decis. Support Syst. **56**, 361–370 (2013)
50. Henderson, R., Divett, M.J.: Perceived usefulness, ease of use and electronic supermarket use. Int. J. Hum. Comput. Stud. **59**, 383–395 (2003)
51. Masrom, M.: Technology acceptance model and e-learning. Technology **21**, 81 (2007)
52. Yang, H., Lee, H., Zo, H.: User acceptance of smart home services: an extension of the theory of planned behavior. Ind. Manag. Data Syst. (2017)
53. AL-Rasheedi, M., et al.: Public awareness of coronavirus (COVID-2019) in Qassim Region Saudi Arabia (2020)
54. Nunnally, J.C., Bernstein, I.H.: Psychometric theory, United States (1994)
55. Fornell, C., Larcker, D.F.: Structural equation models with unobservable variables and measurement error: algebra and statistics. J. Mark. Res. **18**(3), 382–388 (1981)
56. Chin, W.W.: The partial least squares approach to structural equation modeling. Mod. Methods Bus. Res. **295**, 295–336 (1998)
57. Hair Jr, J.F., Anderson, R.E., Tatham, R.L., Black, W.C.: Multivariate data analysis with readings (1995)
58. Henseler, J., Ringle, C.M., Sarstedt, M.: A new criterion for assessing discriminant validity in variance-based structural equation modeling. J. Acad. Mark. Sci. **43**(1), 115–135 (2014). https://doi.org/10.1007/s11747-014-0403-8
59. Sarstedt, M., Henseler, J., Ringle, C.: Multi-group analysis in partial least squares (PLS) path modeling: alternative methods and empirical results. Adv. Int. Mark. **22**, 195–218 (2011)

Conceptual Modeling for Enterprise Systems

Towards Data Ecosystem Based Winter Road Maintenance ERP System

Līva Deksne[1], Jānis Grabis[1(✉)], and Edžus Žeiris[2]

[1] Institute of Information Technology, Riga Technical University, Kalku 1, Riga 1658, Latvia
{liva.deksne,grabis}@rtu.lv
[2] ZZ Dats, Elizabetes iela 41/43, Riga 1010, Latvia
edzus.zeiris@zzdats.lv

Abstract. Traditional Enterprise Resource Planning (ERP) systems are geared towards processing organizations' internal information. However, business processes supported by these ERP systems increasing rely on external data. In order to understand these data dependencies, it is proposed to use data ecosystem models instead of traditional ERP implementation reference models to deploy ERP systems for their users. The winter road maintenance case is explored as a representative case of ERP systems using variety of data sources and requiring advanced data processing features. The winter road maintenance data ecosystem model is developed by means of participative modeling and application of the ecosystem model to configure ERP systems for their users is outlined. The data ecosystem approach allows organizations to partners providing data and data processing services for implementation and operation of the system.

Keywords: ERP systems · Data ecosystem · Winter road maintenance

1 Introduction

Enterprise Resource Planning (ERP) systems are used to plan and to record business activities. The recorded data are further used for reporting and business analysis. Winter road maintenance is one of the typical functions of municipalities or similar responsible organizations [4]. The work effort, material consumption and work results should be booked in the resource planning system for the winter road maintenance case. Various data sources are needed to record and to analyze this information [10]. The centralized data management is complicated because the winter road maintenance involves variety of actors [15] and the data sources are temporal and geographically dispersed. Therefore, users of the ERP system cannot be sure about available data sources and there might be complex data interdependences.

This paper proposes to address this problem by explicitly mapping the winter road maintenance data ecosystem and using this information to setup the ERP system. The data ecosystem is a complex network of organizations that exchange and use data, information and knowledge [13]. In this paper, three central players in the data ecosystem are organizations tasked with winter road maintenance (thus, needing suitable ERP systems),

© Springer Nature Switzerland AG 2021
R. A. Buchmann et al. (Eds.): BIR 2021, LNBIP 430, pp. 69–83, 2021.
https://doi.org/10.1007/978-3-030-87205-2_5

IT consultants providing the ERP system and parties providing or using road mainte-nance data. The IT consultants provide the winter road maintenance ERP system to the responsible organizations. In order to adopt the packaged ERP solution, the responsible organization needs to understand data availability and interdependences. It is assumed that information about the parties providing and using road maintenance information is represented in the data ecosystem model. The data ecosystem model is used by the IT consultants to setup the ERP model for the responsible organization taking into account its specific information needs and data supply constraints.

The objective of this paper is to propose an approach for data ecosystem model based configuration of winter road maintenance ERP system. The data ecosystem model is developed in interactive modeling sessions with stakeholders involved in winter road maintenance processes. The Capability Driven Modeling (CDD) method [19] is used to create the data ecosystem model. The advantage of using the CDD method is integration of strategic level modeling with operational data level modeling and analysis as well as explicit representation of data value chain. The data ecosystem model describes all data items relevant to winter road maintenance and their ownership relationships. It is used as a reference model to create model instances for organization implementing the ERP system. Model instances are created on the basis of the data ecosystem model and they are used to configure the ERP system. The contributions of the paper are development of the empirically based winter road maintenance data ecosystem model and application of this model as a reference model for implementation of the ERP systems.

The rest of the paper is organized as follows. Section 2 describes the case motivating the proposed and research and gives a brief overview of related work. The research method is described in Sect. 3. The winter road maintenance data ecosystem model is presented in Sect. 4 and its application is explored in Sect. 5. Section 6 concludes.

2 Background

2.1 Motivational Case

This research is carried out as a part of the industry sponsored project on development of an integrated ERP system for winter road maintenance. The IT consulting company provides a specialized ERP system to municipalities in Latvia. The system consists of various services including financial accounting, assets management, social welfare services and many other services. In order to expand in the area of data driven services including sensing technologies, the winter road maintenance case was selected as a pilot for developing modules of the ERP system with advanced data analytical functionality.

Winter road maintenance is performed at state or regional level. This research initially focuses on the regional level, where municipalities are tasked with road maintenance and they contract service providers to maintain the roads. Both parties need to exchange information to plan, report, control and analyze the winter road maintenance operations. These operations strongly depend on external factors [4] to obtain data about road con-ditions and a number of parties including citizens, public transportation and emergency services need up-date information on road conditions and maintenance work. Thus, the case is characterized by rich information flows among the parties involved.

The existing research on winter road maintenance and functionality of available information systems were analyzed. A summary of the studies is included in Sect. 2.2. Several theoretical and practical challenges were identified:

- inadequate exchange of data and knowledge between the data providers and consumers.
- insufficient coverage as road information systems do not provide full picture of the road conditions on regional and national scale.
- road maintenance company workers are unable to access video streams from several stationary cameras since they might contain personally identifiable information.
- traditional analytical models are unable to use unconventional data sources such as Waze alerts or information about detected snowy road.
- available winter road maintenance systems are closed systems, which are not intended for integrating unconventional data sources and third-party vendor delivered decision support and data interpretation modules and adjustment modules.
- it is complicated to measure the effectiveness of winter road maintenance operations and rationality of decisions made because that requires wide coverage, use of unconventional data sources and regional platform tailoring.
- winter road maintenance systems are not tailored to suit regional peculiarities including data availability considerations.

2.2 Related Work

Winter road maintenance concerns ensuring safe driving conditions in winter weather conditions with focus on snow removal and deicing. The problem is characterized by high degree of complexity due to information processing and decision-making challenges. Perier et al. [15] provides an extensive review of decision-making algorithms used in winter road maintenance. That includes design of plowing and removing operations, vehicle routing, depot location and resource allocation. From the information processing perspective, variety of factors need to be taken into account for planning and execution of the winter road maintenance operations. Planning support is often provided by forecasting modules [1, 16]. A Pisano et al. [18] propose functional prototype of winter road maintenance decision support system. Prototype is focused on forecasting and data fusion techniques merging them with winter road maintenance rules of practice to provide forecast of road surface conditions and treatment recommendations. The prototype is based on data retrieval from National Weather Service and supplemental weather prediction models, additional road condition and treatment model is used for data ingest and forecast output generation. Hinkka et al. [7] analyzes existing information systems for winter road maintenance and calls for integrated solutions with emphasis on collaboration among various actors involved in winter road maintenance. They point out that the current systems focus on the centralized data processing model what leads to local optimization of maintenance decisions. The most significant future research problems are sharing of maintenance data with road users, selection of appropriate maintenance actions and usage of alternative data collection mechanisms.

Decision-making in winter road maintenance is a difficult task, as various data need to be processed to assess the current situation. The main goal of this complex process is to decide what action to take in certain weather and road conditions. As Nixon [12] has described, *decision support systems intend to support the decision being made, not to make the decision. However, in WRM various systems suggest specific actions that in general are directly applicable.* This is basically achieved defining and using sets of rules. Nixon [12] study presents a simplified schematic representation of the information flows involved in winter road maintenance that consists of two main modules – (1) information input module and (2) action and decision module. Information input module ingests and pre-process data to gain information, but action and decision module provides initial and subsequent evaluation in order to apply recommended action on roads.

The winter road maintenance system should support integration of data from various sources including national weather service, cameras, mobile weather stations and usage of intelligent data processing [9]. Kociánová [9] has designed road weather information system (RWIS) that is based on monitoring and forecasting of weather and road surface conditions in Slovak Republic. System provides data-driven decision making for effective and timely winter road maintenance activities. There are three main blocks used in RWIS – (1) current data module, that receives and processes data from multi-type data sources, (2) road weather forecast module, that uses forecast data from National Weather Service, commercial provider and special computational models, (3) treatment recommendation module. At the national level, winter road maintenance system can be implemented as a centralized solution. Web services allow to create reusable solutions adapted for specific clients [3]. A prototype of the system for predicting road conditions along the anticipated travel route is developed.

The data ecosystem approach to the maintenance activities has been proposed [17]. The data ecosystem includes such players as equipment manufacturers, data providers, service providers, information service providers as well as users. Value chain analysis is applied to describe and analyze existing data ecosystems. Value creation between many involved parties is defined. Method is used to analyze the role of stakeholders in the winter road maintenance data ecosystem, without focusing on the identification of information retrieval mechanisms in service delivery to use a data ecosystem as a basis for system development. A stakeholder mapping approach also has been applied to determine requirements towards road maintenance information systems [2]. The focus in this work is on equipment operator support.

The aforementioned intelligent winter road maintenance systems focus on road monitoring and clearing while organizations need integrated solutions for all winter road management processes including reporting, controlling and accounting. Fu et al. [5] define winter road maintenance scheduling problem to develop an operations plan for the available service vehicles that specify route assignment, service type and the corresponding start time. Adaptive weather information system for surface transportation Pikalert introduced in Siems-Andersons et al. [20] study ingests weather and vehicle data and performs quality control using algorithms in order to set confidence values to the data and enable filtering options for use in the system. Different modules are used that include information visualization, alerts, forecasts, as well as decision support system. Three main blocks are used in system design that are based on data flow. Decision support systems and their components in different studies are summarized in Table 1. Yu et al. [23] points out that proactive asset management, heterogeneous data processing, business collaboration, intelligent decision-making support and comprehensive visualization and supervision are needed. These solutions are increasingly used in infrastructure management [11, 22].

Table 1. Decision support system, architecture components

No	Architecture components	Study
1	Current Data Module Road Weather Forecast Module Treatment Recommendation Module	Kociánová [9]
2	System backend (Data ingest, vehicle data translator, road condition treatment, road weather forecast system, road weather hazard assessment, road weather alert) Enhanced Maintenance Decision Support System (backend and display) Motorist Advisory Warning System (backend, web display and amobile app)	Siems-Andersons et al. [20]
3	Data collection module Road temperature and state forecast module Precipitation forecast module Decision-making module Treatment recommendations	Bazlova et al. [1]
4	Road condition and treatment module Supplemental weather models Data Fusion – Road Weather Forecast System	Pisano et al. [18]
5	Information input module Action and decision module	Nixon [12]
6	Data ingest and processing module Road Weather Forecast System Forecast and Observation Data Store Road condition and treatment module Frost Module Alert Generator System Display	Petty et al. [16]

Implementation of complex ERP systems is time-consuming and knowledge intensive. That can be addressed by reusing existing knowledge in a form of reference models to streamline ERP implementation [14]. However, the reference models rarely support the data ecosystem view of implementation of complex ERP systems.

3 Method

Conceptual modeling method is chosen considering its ability to focus on defining system services based on business goals, while being able to form a structured information retrieval from multi-source data provided by many parties in a volatile environment. The CDD method and its extension for modeling data ecosystems are used to elaborate the data ecosystem model [6]. The model is developed using the corresponding meta-model. The participatory modeling is used to create the data ecosystem model. The CDD approach allows to create a complex data ecosystem that integrates different stakeholder perspectives and supports flexibility for data users and enables data-driven decision-making. The CDD provides an opportunity to identify an organization's needs without considering the technical details of how they will be achieved. This approach makes it possible to increase flexibility needs and offers methodological support for the design and delivery of business solutions [19]. A data ecosystem model based on the CDD approach, which tolerates different types of data sources from stakeholders, is an appropriate method to identify the contextual elements from data and the adaptations needed to achieve business goals. This factor is particularly important in the event of future changes to the data ecosystem, as well as the emergence of new data sources, as the system can be adapted without losing established business capabilities, while providing support for a volatile environment. The CDD is an appropriate methodology for the data ecosystem, as it has the ability to provide different levels of model abstraction, tool support for model development, data processing execution, harmonization of data users and providers, and general customization solutions for individual ecosystem actors and knowledge sharing [8].

3.1 Meta-model

The data ecosystem model is developed using the capability-based data ecosystem modeling method, whose meta-model is shown in Fig. 1. Various organizations have capabilities to provide their business services. The capabilities are defined as an ability to deliver business services in volatile environments. A data ecosystem is a complex network of organizations collaborating and interacting during the capability delivery. There are various types of interactions, and this research focuses on data and information exchange type of interactions. The meta-model assumes that organizations referred as to parties operate in the data ecosystem. They possess capabilities to achieve their business goals measured by Key Performance Indicators (KPI). Successful capability delivery depends on other parties in the data ecosystem. Thus, evolution of the data ecosystem contributes to development of capabilities, and this development spurs further evolvement of the data ecosystem.

Although there are various types of participants involved in data ecosystems, only providers and consumers are distinguished. The data ecosystem brings together parties on a premise that the consumers need specific assets to deliver capabilities. and the providers possess these assets and are willing to share them subject to certain condition. The asset is a resource contributing to delivery of capabilities. In the data ecosystem, key assets are data and their processing abilities. Therefore, the foundational elements of the CDD method, namely, measurable property, context element and adjustment are considered as assets. The assets are made available for capability delivery as services. The services are software components providing a specified component in response to the consumer request. The capability delivery information system is developed as a composition of the services. It consists of both traditional services and adaptive and data services. The capability model concerns the adaptive and data services while the traditional services responsible for generic business services (e.g., sales order creation) are modeled and developed using organization's internal engineering processes.

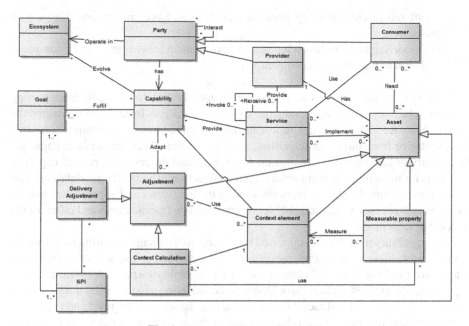

Fig. 1. Data ecosystem meta-model

The capability delivery is data driven, i.e., the capability development environment is actively observed and capability delivery is adapted according to these observations to ensure efficient operations. The context elements are used to describe the relevant environment factors affecting the capability delivery. The context elements assume either categorical or continuous values. The context is measured using measurable properties, which are actual observations of some phenomena. In the data ecosystem model, general measurable properties are represented. The actual providers of measurements are listed in a catalog. During the analysis of the data ecosystem model, the catalog is queried to determine available sources of the required measurable properties.

The adaptation of capability delivery and context processing are specified using adjustments, which are expressed as algorithmic recommendations. The context calculation adjustment transforms measurable properties into context elements with clear business meaning. The delivery adjustment changes the capability delivery in response to changing contextual situations and it uses KPI to steer capability delivery towards its goals.

3.2 Modeling Process

The data ecosystem model is created using the CDD method and its data ecosystem modeling extension. Participatory modeling techniques are used to involve stakeholders in the modeling process [21]. The modeling process was organized to involve different types of stakeholders and to avoid domination of a single perspective on the winter road maintenance. It was performed in three stages:

1. A draft model created using document analysis and preliminary interviews with stakeholders;
2. Joint modeling sessions involving researchers, domain area experts and IT company;
3. Validation of the model with stakeholders.

In the first stage, the focus was on identification of capabilities and their supporting services. The interviews were conducted with representatives from the IT company, the Latvian State Road oversight institution, the largest Latvian road maintenance company, representative from three municipalities, the Latvian State Weather service, Riga (the largest city in Latvia) traffic monitoring department and the researchers with expertise in enterprise modeling and architecture, civil engineering experts. The model developed was discussed with the stakeholders involved in the interviews and appropriate changes were made if necessary. The stakeholders commented on completeness and utility of the data ecosystem model.

The capability modeling is supported by suitable tools though requiring tool-specific skills. To facilitate participatory modeling, it was decided to use a user-friendly and interactive platform Miro. All elements from the capability meta-model were assigned a type-specific colored sticker. Participants of the modeling sessions used the stickers to add items to the model and facilitators ensured that the stickers are used and connected in a consistent manner. The models created in the in-formal modeling environment subsequently can be transformed into formally defined models.

4 Data Ecosystem Modeling

Deployment of the ERP system is guided by a data ecosystem model. The data ecosystem model shows all parties involved in the winter road maintenance and serves to understand and analyze interactions among the parties as well as a reference model used by an IT consulting company to implement the ERP system for a specific customer involved in the data ecosystem.

4.1 Capabilities

The parties involved in the data ecosystem jointly should have capabilities to execute all winter road maintenance business processes. The capability delivery and business process execution are provided by a set of services. These services define the desired functionality of the resource planning system.

To establish comprehensive scope of the ERP system, key required capabilities are identified at the data ecosystem level as well as services supporting these capabilities (Fig. 2). Individual members of the data ecosystem do not necessarily need all capabilities and services.

The Road clearing capability is an ability to remove snow and ice from the road surface according to pre-defined service level requirements. It is supported by sequential services of planning the road maintenance work, work tracking and reporting the completed work. The key feature of this capability is a need for data integration among parties monitoring, controlling and executing the maintenance work. The Road conditions monitoring capability is an ability to timely determine road conditions and required maintenance activities as well as to notify the concerned parties. The supporting services are responsible for data gathering and interpretation what in turn serves as an input to services creating work orders and disseminating road conditions related alerts. These services are characterized by variety of data provision and dissemination channels. The Road maintenance controlling capability is an ability to inspect and approve the work performed. It also includes providing feedback to citizens and other parties the maintenance work completion especially with regards to notifications received. The Performance management and Environment impact assessment capabilities concern abilities to analyze efficiency of the maintenance work. Finally, the Road maintenance accounting capability is ability to book road maintenance effort and resource consumption as well to invoice and pay for the maintenance services.

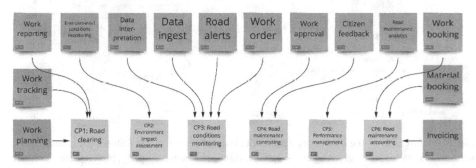

Fig. 2. Winter road maintenance capabilities (yellow stickers) and services (green stickers). (All arrows represent the same relationship – provide). (Color figure online)

4.2 Data Ecosystem Model

The data ecosystem model is created by further elaboration of capabilities and their consumer/provider relationships. The parties involved in the winter road maintenance data ecosystem are: 1) regional road authorities (municipalities), 2) state road authorities, 3) winter road maintenance service providers, 4) IT consulting companies (providing ERP systems), 5) road and weather information providers, 6) web based information systems, 7) various observations' data providers and 8) information dissemination outlets. Figure 3 shows a fragment of the data ecosystem model centered around the Road conditions monitoring capability. The capability requires three services including Road monitoring service, Work Order and Driver notification services. The Smart sign management service is an external service provided by the Road governance company. This service is needed if driver notifications are pushed to roadside signs. The model also shows that the Driver notification service is consumed by the external traffic web service.

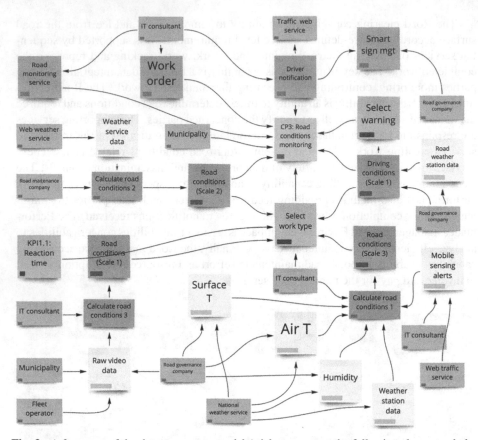

Fig. 3. A fragment of the data ecosystem model (sticker represent the following elements: dark yellow – capability, dark green square – service, light green square – party, pink square – context element, pink rectangle – context calculation, purple – adjustment, blue – KPI, light yellow – measurable property; arrows represent different relationships between elements based on the data ecosystem meta-model in Fig. 1). (Color figure online)

The capability delivery goals are identified for every capability along with KPI used to measure the goals. For instance, the primary goal of the Road clearing capability is to comply with regulatory requirements what is measured by the road maintenance execution time. Similarly, contextual factors affecting the capability delivery as well as ways to measure them are identified. There are context elements characterizing road conditions and driving conditions. The former describes observations about road surface and there are different scales to characterize them. For instance, Road conditions (Scale 1) include values such as Bare, Partly Covered and Covered. The Driving conditions characterize perceived easy of driving and has values Good, Fair, Caution and Poor. These values are determined using context calculations, which include analytical calculations, rules and machine learning. The context calculations are assets provided by parties in the data ecosystem. Measurable properties are used to calculate the context elements. There are measurable properties like Air temperature having several providers. Similarly, the Surface temperature is provided by several parties though they have different geographical coverage.

There are two adjustments Select work type and Select warning. The adjustments describe capability delivery adaptation logics and represent intelligent data processing. The Select work type uses the Road conditions context and the Reaction time KPI to decide type of clearing and serves as input to the Work order service. Similarly, the Select warning adjustment uses a set of rules to determine the type of notifications to disseminate.

The created data ecosystem model consists of high-level elements that do not identify specific methods for their implementation in the system. It identifies services, data providers and the use of their assets to provide capabilities, with an emphasis on creating context and exchanging information between many parties in order to achieve defined business goals. Each element created and its main function is described separately in the documentation. This prevents misinterpretation during the system development phase, as only general notations are included in the model visualization.

The data ecosystem model can be used for different purposes including analysis of interactions among the parties and viability of the data ecosystem as well configuration of the ERP system for individual parties in the data ecosystem.

5 Application

An example is used to demonstrate application of the data ecosystem model for configuration of the ERP system.

5.1 Model Instance

An instance of the data ecosystem model is created for the Smallcity municipality, which was also involved in development of the data ecosystem model. The municipality is small town in central Latvian with approximately 30 thousands inhabitants and served by a private contractor. The model instance represents winter road maintenance concerns from the perspective of this municipality (Fig. 4). To create the model instance, the municipality specifies its goals and required capabilities in the data ecosystem model. Selection of the capabilities identifies required abilities of the municipality itself while selection of goals points to capabilities needed from other parties. The capabilities selected lead to services and assets the municipality needs. In this case, the municipality opted for the ability to create work order to address emergency situations and to ensure quick resolution. It did not select Driver notification service and will use two different ways to determine road conditions and to select an appropriate work type. The raw video data captured by municipality operated network of monitoring cameras is used to identify snow covered roads. The IT consulting company owns assets for machine learning based analysis of the video data. The data sources provided by the National weather service and a mobile application is used to determine road conditions on Scale 3.

The model instance shows that the Small city municipality needs to engage with the IT consulting company and National weather service to deliver the capability. It also specifies the information sources needed what includes weather station data, mobile sensing alerts and raw video data. The data ecosystem model shows that the mobile sensing alerts also are provided by a web traffic service. However, a contract between the municipality and the web traffic service was needed and the municipality decided to forgo using this asset. The intelligent data processing functionality required includes calculation of road conditions and selection of work type. The IT consulting company has these assets and these could be used as a part of the ERP system. This particular fragment shows that two services of the ERP system need to be enabled for the Smallcity municipality.

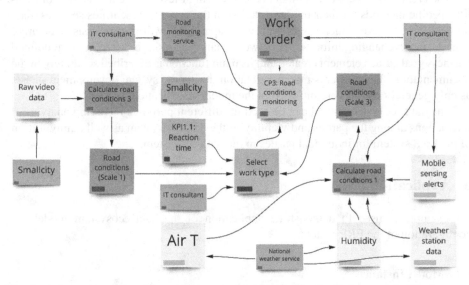

Fig. 4. A fragment of the data ecosystem model instance for Smallcity municipality (arrows represent different relationships between elements based on the data ecosystem meta-model in Fig. 1)

5.2 System Configuration

The model instance is envisioned to be used to configure the ERP system implemented for a particular organization or user (Fig. 5). The out-of-box ERP system supports capabilities defined in the data ecosystem model. In order to implement the ERP system for a particular user (e.g., municipality), the user specific model instance is created as described in Sect. 5.1. Similarly, an instance of the ERP system is also created for the user. The ERP system's instance is configured to include services required for capabilities included in the model instance. The data ecosystem model also specifies, which data sources and data processing assets and services implementing these assets are needed.

The model instance fragment shows that the Road monitoring and Work order services need to be enabled in the ERP system. The Raw video data, Weather station data (including Air T and Humidity) and Mobile sensing alerts need to be connected to the ERP system. The Select work type, Calculate road conditions 1 and Calculate road conditions 2 assets and their implementing services are also needed.

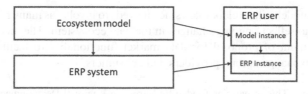

Fig. 5. The data ecosystem based configuration of ERP system

6 Conclusion

The data ecosystem model-based approach to configure ERP systems for specific users is proposed. That allows to understand interactions between company using the ERP system and various external parties providing data and data processing assets. The data ecosystem model also allows sharing domain knowledge among the parties involved. It was validated by domain experts and stakeholders.

Close collaboration with business representatives is necessary to develop a data ecosystem model based on CDD approach. Capabilities are defined considering business goals and KPIs, and because their identification is a foundation of the data ecosystem, poor collaboration with business representatives in a given field can result in an inefficient data ecosystem from a business perspective. The created data ecosystem focuses on capability identification in case of many data providers. The method used allows to construct information retrieval mechanisms, pointing out their logical structure in capability delivery, while clearly linking data providers and data interdependencies.

Also, model validation by business representatives is a necessary step. Model elements are created based on modeler's assumptions, therefore the validation by business representatives is important to validate model's suitability to business needs.

Identification of multi-type data and their processing methods in context elements in order to gain valuable information strengthens the resilience of data ecosystem. The ability to add new data providers and to use their data processing assets will require fewer changes to the data ecosystem in the future if different types of data are identified at the first data ecosystem model development. Resilience of data ecosystem is highly dependent on data providers. Used method allows to calculate and monitor their importance in each specific context delivery, which is very crucial for risk management and can be used by data consumers to determine their dependence on other parties.

The winter road maintenance module will be developed as an additional component of the IT company's ERP system. It will contain an analytical component and a transactional component. The analytical component will primarily address needs of the Road conditions monitoring, Environment impact assessment and Performance management capabilities while the transactional component will address needs of the Road Clearing, Road maintenance controlling and Road maintenance accounting capabilities. The setup of the analytical component will particularly benefit from using the data ecosystem mode to determine the necessary data sources and data processing assets. Although the current research centers around the ERP system of a particular vendor, it is envisioned that different services can be provided by other IT consulting companies as well.

The data ecosystem model will be further elaborated during development of the ERP system and its piloting jointly with municipalities. The mechanisms for maintaining the data ecosystem model will be elaborated since it needs to involve as many relevant parties as possible and to reflect ongoing changes in the data ecosystem. The mechanisms will include development of proposal for data market functionality to facilitate mutually beneficial data exchange among providers and consumers.

Acknowledgement. This research is funded by European Regional Development Fund Project Nr. 1.1.1.1/19/A/003 "IWiRoM: Development of a new type of Intelligent Winter Road Maintenance information system and ERP integration solution for improving efficiency of maintenance processes" Specific Objective 1.1.1 "Improve research and innovation capacity and the ability of Latvian research institutions to attract external funding, by investing in human capital and infrastructure" 1.1.1.1. measure "Support for applied research" (round No. 3).

References

1. Bazlova, T., Bocharnikov, N., Pugachev, A., Solonin, A.: Decision support system for winter maintenance-research and practice. In: Proceedings of SIRWEC 16th International Road Weather Conference, Helsinki, Finland, pp. 23–25 (2012). http://sirwec.org/wp-content/upl oads/Papers/2012-Helsinki/59.pdf
2. Campos, J., Kans, M., Håkansson, L.: Information system requirements elicitation for gravel road maintenance – a stakeholder mapping approach. In: Ball, A., Gelman, L., Rao, B.K.N. (eds.) Advances in Asset Management and Condition Monitoring. SIST, vol. 166, pp. 377–387. Springer, Cham (2020). https://doi.org/10.1007/978-3-030-57745-2_32
3. Casselgren, J., Bodin, U.: Reusable road condition information system for traffic safety and targeted maintenance. IET Intell. Transp. Syst. **11**(4), 230–238 (2017). https://doi.org/10.1049/iet-its.2016.0122
4. Dey, K.C., et al.: Potential of intelligent transportation systems in mitigating adverse weather impacts on road mobility: a review. IEEE Trans. Intell. Transp. Syst. **16**(3), 1107–1119 (2015). https://doi.org/10.1109/TITS.2014.2371455
5. Fu, L., Trudel, M., Kim, V.: Optimizing winter road maintenance operations under real-time information. Eur. J. Oper. Res. **196**(1), 332–341 (2009). https://doi.org/10.1016/j.ejor.2008.03.001. ISSN: 0377-2217
6. Grabis, J., et al.: Capability management in resilient ICT supply chain ecosystems. In: ICEIS 2020 - Proceedings of the 22nd International Conference on Enterprise Information Systems, pp. 393–400 (2020)
7. Hinkka, V., et al.: Integrated winter road maintenance management - new directions for cold regions research. Cold Reg. Sci. Technol. **121**, 108–117 (2016). https://doi.org/10.1016/j.coldregions.2015.10.014
8. Kampars, J., Zdravkovic, J., Stirna, J., Grabis, J.: Extending organizational capabilities with Open Data to support sustainable and dynamic business ecosystems. Softw. Syst. Model. **19**(2), 371–398 (2019). https://doi.org/10.1007/s10270-019-00756-7
9. Kociánová, A.: The intelligent winter road maintenance management in Slovak conditions. Procedia Eng. **111**, 410–419 (2015). https://doi.org/10.1016/j.proeng.2015.07.109. TFoCE
10. Mahoney III, W.P., Myers, W.L.: Predicting weather and road conditions: integrated decision-support tool for winter road-maintenance operations. Transp. Res. Rec. **1824**, 98–105 (2003). https://doi.org/10.3141/1824-11
11. Michele, D.S., Daniela, L.: Decision-support tools for municipal infrastructure maintenance management. Procedia Comput. Sci. **3**, 36–41 (2011). https://doi.org/10.1016/j.procs.2010.12.007

12. Nixon, W.A.: Using a maintenance decision support system in winter service operations. J. Public Works Infrastruct. **2**(1), 74–84 (2009)
13. Oliveira, M.I.S., Lóscio, B.F.: What is a data ecosystem? In: ACM International Conference Proceedings Series (2018). https://doi.org/10.1145/3209281.3209335
14. Pajk, D., Indihar-Štemberger, M., Kovačič, A.: Enterprise resource planning (ERP) systems: use of reference models. In: Grabis, J., Kirikova, M. (eds.) BIR 2011. LNBIP, vol. 90, pp. 178–189. Springer, Heidelberg (2011). https://doi.org/10.1007/978-3-642-24511-4_14
15. Perrier, N., et al.: A survey of models and algorithms for winter road maintenance. Part I: system design for spreading and plowing. Comput. Oper. Res. **33**(1), 209–238 (2006). https://doi.org/10.1016/j.cor.2004.07.006
16. Petty, R.K., Mahoney III, W.P., Cowie, J.R., Dumont, A.P., Myers, W.L.: Providing winter road maintenance guidance, an update of the federal highway administration maintenance decision support system. In: Transportation Research Circular, Surface Transportation Weather and Snow Removal and Ice Control Technology, pp. 199–214. Transportation Research Board (2008). https://doi.org/10.17226/17626
17. Pilli-Sihvola, E., et al.: Evolving winter road maintenance ecosystems in Finland and Hokkaido, Japan. IET Intell. Transp. Syst. **9**(6), 633–638 (2015)
18. Pisano, P.A., Stern, A.D., Mahoney III, W.P., Myers, W.L., Burkheimer, D.: Winter road maintenance decision support system project: overview and status. In: Transportation Research Circular (2004). Paper number: E-C063
19. Sandkuhl, K., Stirna, J. (eds.): Capability Management in Digital Enterprises. Springer, Cham (2018). https://doi.org/10.1007/978-3-319-90424-5
20. Siems-Anderson, A.R., Walker, C.L., Wiener, G., Mahoney, W.P., Haupt, S.E.: An adaptive big data weather system for surface transportation. Transp. Res. Interdisc. Perspect. **3**, 100071 (2019). https://doi.org/10.1016/j.trip.2019.100071. ISSN: 2590-1982
21. Stirna, J., Persson, A., Sandkuhl, K.: Participative enterprise modeling: experiences and recommendations. In: Krogstie, J., Opdahl, A., Sindre, G. (eds.) CAiSE 2007. LNCS, vol. 4495, pp. 546–560. Springer, Heidelberg (2007). https://doi.org/10.1007/978-3-540-72988-4_38
22. Wei, L., et al.: A decision support system for urban infrastructure inter-asset management employing domain ontologies and qualitative uncertainty-based reasoning. Expert Syst. Appl. **158**, 113461 (2020). https://doi.org/10.1016/j.eswa.2020.113461
23. Yu, G., et al.: RIOMS: an intelligent system for operation and maintenance of urban roads using spatio-temporal data in smart cities. Future Gener. Comput. Syst. **115**, 583–609 (2021). https://doi.org/10.1016/j.future.2020.09.010

Trends on the Usage of BPMN 2.0 from Publicly Available Repositories

Ivan Compagnucci⬭, Flavio Corradini⬭, Fabrizio Fornari(✉)⬭,
and Barbara Re⬭

University of Camerino, Via Madonna delle Carceri 7, Camerino 62032, Italy
{ivan.compagnucci,flavio.corradini,fabrizio.fornari,barbara.re}@unicam.it

Abstract. Business Process Model and Notation is the de facto standard for graphically modelling business processes. Since its first release in 2004, it evolved until reaching the actual 2.0 version, which presents more than 85 elements. Despite the notation being rich in graphical elements, initial studies show that only a subset of the BPMN elements is actually used. This paper aims at investigating whether the BPMN vocabulary adopted nowadays by model designers shows some particular trends. We collected 25,590 models from six online repositories to conduct such an investigation, and we analysed them. We report and discuss the obtained results providing insights on the correlations in the BPMN vocabulary and the resulting complexity of BPMN models.

Keywords: BPMN · Business process modelling · Models repositories

1 Introduction

Business Process Model and Notation[1] (BPMN) is an OMG standard [19] which provides a graphical notation for the modelling of business processes. The notation emerged as the de facto standard to support business process modelling for all the business stakeholders (e.g., business analysts who create and refine the processes, technical developers who implement them, business managers who monitor and manage them). The success of BPMN comes from its versatility and capability to represent business processes for different purposes. The intuitive graphical representation of more than 85 elements made BPMN widely accepted by both the industry and the academia.[2] Thanks to its intuitiveness, BPMN can be easily used to design processes and their interactions. These models can be used to communicate and interchange the business requirements of a business process and provide the underpinning of the actual process implementation. Therefore, a BPMN model can be understood by all stakeholders involved in the

---

[1] BPMN 2.0.2 is the latest version released in December 2013 https://www.omg.org/spec/BPMN/2.0.2 formally published by ISO as the 2013 edition standard: ISO/IEC 19510.

[2] More than 70 tools support BPMN (http://www.bpmn.org).

© Springer Nature Switzerland AG 2021
R. A. Buchmann et al. (Eds.): BIR 2021, LNBIP 430, pp. 84–99, 2021.
https://doi.org/10.1007/978-3-030-87205-2_6

process. Since its first release (May 2004), BPMN has been also protagonist of several research contributions ranging from studies related to: the usage of the notation (e.g., [1,3,10,18,21]), the definition of a rigorous semantics for each notation element (e.g., [6,8,16,22]), the evaluation of BPMN models qualities (e.g., [2,5,11,12]), the definition of notation extensions to incorporate specific application domain aspects (e.g., [4,23]), and many others.

This paper reports the results of an investigation on whether the BPMN vocabulary adopted nowadays by model designers shows some particular trends. We collected a total of 25,590 models from six collections, some of which have already been used for conducting research activities. We focused on studying the overall usage of modelling elements and their correlation and combined usage in designing a business process model. We also investigated the complexity of the notation, highlighting the actual difference between the BPMN practical and the theoretical complexity. From the conducted study, we conclude whether or not the usage of the BPMN notation for modelling business processes has changed during the years, especially comparing our results with previous studies conducted on the topic [18].

The paper is organised as follows: Sect. 2 shows the methodology we used for harvesting BPMN models. Section 3 reports the overall usage of BPMN notation. Section 4 reports about model complexity in terms of model size and variety of elements used. Section 5 highlights the correlation in modelling BPMN elements in combination and also the set of most popular vocabulary subsets of BPMN. A comparison with related works is presented in Sect. 6. Finally, Sect. 7 discusses limitations and future work.

2   Models Harvesting

For conducting our analysis, we gathered different repositories of business process models designed with the BPMN notation. In particular, we refer to six repository such as: BIT process library[3], Camunda BPMN[4], eCH-BPM, Gen-MyModel, GitHub, and RePROSitory.

The *BIT process library*, is composed by 850 BPMN files containing abstract business process used by IBM WebSphere Business Modeler V6 and V7. Those models have been made available by IBM for the practical validation of the soundness-checking approaches and tools [9]. The *Camunda BPMN* collection stores a total amount of 3,739 unique BPMN models designed in BPMN which have been made available, on GitHub[5], by the CAMUNDA company. These diagrams have been designed in BPMN training sessions, which have been given since 2008 until 2015 when the models have been released. The *eCH-BPM* model collections includes 117 models downloaded from the eCH process platform[6]. The published process models are examples of Swiss public administration's processes

---

[3] The full collection name is "IBM Research GmbH, BIT process library, release 2009".
[4] The full collection name is "Camunda BPMN for Research".
[5] https://github.com/camunda/bpmn-for-research.
[6] http://www.ech-bpm.ch/de/process-library.

that, have been designed by the eCH association, in cooperation with municipalities, cantons and federal agencies as well as the BPM specialist community. The *GenMyModel* collection refers to 11,460 models that have been downloaded, at the time of writing, from the GenMyModel platform[7]. Users can experiment with the platform's functionalities provided that the designed models will be made publicly available for reuse. The *GitHub* collection of models actually consists of a set of 17,203 BPMN models publicly available on GitHub that have been retrieved by Heinze et al. with a procedure described in [14]. The *RePROSitory* collection consists of 560 models that have been harvested, from the proceedings of the BPM conference and manually re-designed with the objective of defining a benchmark of models with a solid literature background. The models are available on the RePROSitory[8] dedicated platform [7].

For each collection of models we run a filtering procedure for removing models with a total number of element less than eight. We derived this threshold value by analysing each models collection and manually inspecting all the models which size was included in a range of zero to ten elements. We concluded that models with a number of elements less than eight were incomplete models that could have compromised the validity of our study. Table 1 reports the amount of models considered from each collection after the application of the filtering procedure. At the end of our models harvesting procedure, we gathered a total of 25,590 models.

Table 1. Models collections overview

Models repositories	Number of considered models	Source
BIT process library	804	Literature
Camunda BPMN	3 721	Training sessions
eCH-BPM	117	Government
GenMyModel	11 156	Mixed
GitHub	9 232	Mixed
RePROSitory	560	Literature

For extracting data from the harvested models, we developed a python script which allows to count the occurrences in a *.bpmn* file of 85 BPMN elements. We based the development of the python script on the tags present in the BPMN 2.0 meta-model[9] so to be able to select all the actual BPMN elements plus some of them characterised by the usage of attributes. We ran our script over the 25,590 retrieved models. The source code and all the details about the usage of this script are reported on the PROS Lab GitHub account[10] together with all the extracted data and the data resulting from the performed statistical analysis.

---

[7] https://www.genmymodel.com/.
[8] http://pros.unicam.it/reprository.
[9] https://www.omg.org/spec/BPMN/2.0/About-BPMN/.
[10] https://github.com/PROSLab/BPMN-element-counter.

3   Overall Usage of BPMN Elements

In this section, we present statistical analysis performed over the entire set of models to detect which BPMN elements have been used to design such models and their frequency.

3.1   Distribution of BPMN Elements over Models

Considering the frequency distribution of the individual BPMN elements over the total amount of models we ranked them from the most frequent element to the less frequent. As it can be observed in Fig. 1, the distribution of BPMN elements follows a power-law distribution. In our study we found out that five elements namely *Sequence Flow, End None Event, Start None Event, Task, and Exclusive Gateway* are present in **more than the 50%** of the overall models. In the range **between 50% and 20%** we found 8 elements respectively Pool, User Task, Lane, Parallel Gateway, Message Flow, Start Message Event, Association and Service Task. **Between 20% and 10%** we found 7 elements respectively *Intermediate Catch Timer Event, Intermediate Catch Message Event, End Terminate Event, Event Based Gateway, Collapsed Sub Process, Conditional Event, Text Annotation.* The remaining elements of the notation (65 of 85) are present in less than 10% of models. **Between 10% and 1%** we found 25 elements mainly related to typed tasks, some particular type of events, data related elements (*Data Object* and *Message*), grouping elements (*Group*), *Expanded Sub-Processes* and *Default Flow*. 36 elements over 85 are present in less than 1% of the models, they mainly include particular types of Activities (*Transaction, Ad Hoc Sub Processes, Task Loop Activity,* etc.), particular types of Events (*Intermediate Catch Multiple Event, Intermediate Throw Multiple Event, etc.*) and other elements such as *Choreography Task, Conversation,* etc. To deepen our analysis, we also derived and reported in Fig. 2 the average number of occurrences of a given construct in a model. The reported values are only related to models that present such a construct. From Fig. 1 we notice that 99.65% of the models include a Sequence Flow (ID 1),[11], while from Fig. 2 we notice such a percentage of models presenting around 16 sequence flows on average. Some peaks can be seen in Fig. 2 even in correspondence of elements that are not present in many models. It is the case of Choreography Participant (ID 72). It is widespread in 0.07% of the models.

3.2   Frequency Distribution of BPMN Elements

We also analysed the frequency distribution of BPMN elements concerning the total amount of elements present in the entire collection of models. In particular, the 25,590 models are composed by 1,038,084 BPMN elements. The frequency

---

[11] As it can be seen from Fig. 1 we associated an ID to each BPMN element so, for presentation purposes, the following analysis and diagrams use the IDs instead of the plain name of the elements.

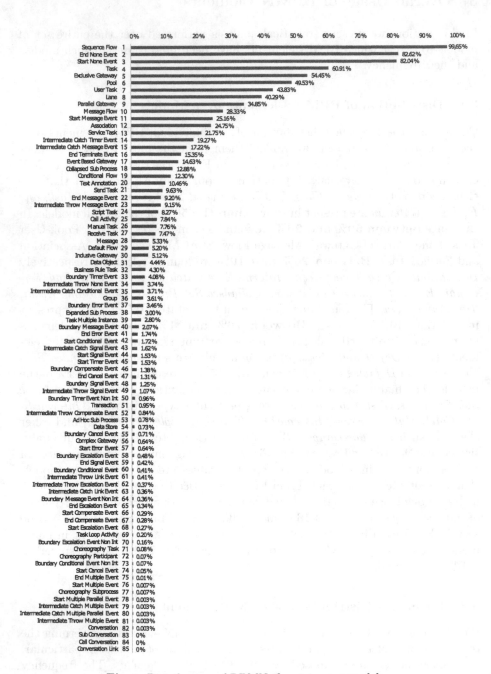

Fig. 1. Distribution of BPMN elements over models

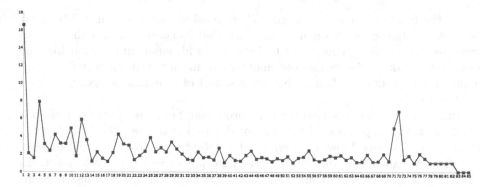

Fig. 2. Average number of occurrence of a given construct in a model

distribution of all BPMN elements is reported in Fig. 3, and it is ordered by following the ranking emerged from Fig. 1. Starting from the left to the right of the figure, we find rank 1, which corresponds to the number of Sequence Flow, rank 2, which corresponds to End None Event, till rank 85, which corresponds to Conversation Link. Besides the frequency distribution of BPMN elements, we also reported the Zipfian distribution (highlighted in red). The Zipfian distribution states that the frequency of words in natural languages is inverse to their rank [24]; this juxtaposition allows us to highlight that BPMN exhibits a very close distribution to that one of word usage in natural languages.

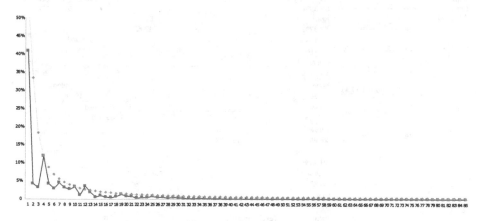

Fig. 3. Frequency plot of BPMN elements by rank

The 85 BPMN elements, according to the standard, can be grouped in eight families: *Activities, Gateways, Events, Connecting Objects, Swimlanes, Artefacts, Choreography* and *Conversation*. Therefore, to provide additional insight on the frequency distribution of BPMN elements, we reported in Fig. 4 statistics over their usage grouped by families; we also distinguished between categories in which the families can be organised. Where possible, we distinguished between basic elements such as Normal Task, Start/End None events, and their *Typed* versions, which allow a modeller to represent additional information like the fact that a task can be carried out manually or in an automatic way. To group additional elements of a family that are not part of a specific category, we used the term *Others*.

In Fig. 4, we can notice that the most prominent family is the one of *Connecting Objects* that represents almost half of the used elements (49.51%), of which mostly are Sequence Flow. Concerning the *Activities* family (22.34%), most of the elements are Tasks (92.39%), more than half of which are Normal Tasks (57.37%); Sub-processes (3.96%) and other (3.65%) activities are less present. About the *Events* family, we notice a predominance of End (41.17%) and Start (35.16%) Events; most of which are of the basic form *None* (respectively 80.52% and 70.03%). The other events following are Intermediate (19.28%) and Boundary (4.38%) events. Intermediate events are divided into Throwing (28.85%) and Catching (71.05%), instead Boundary events are divided into Interrupting (91.26%) and Non-Interrupting (8.74%). Regarding the *Gateways* family, the Exclusive Gateway is the most used (54.63%), followed by the Parallel Gateway (35.37%) and the Others (10%). Elements of the *Swimlanes* family forms the 6.11% of the total and are divided into Pool (47.62%) and the Lanes (52.38%). The *Artefacts* family forms the 1.5% and most of its elements are Text Annotations (53.62%).

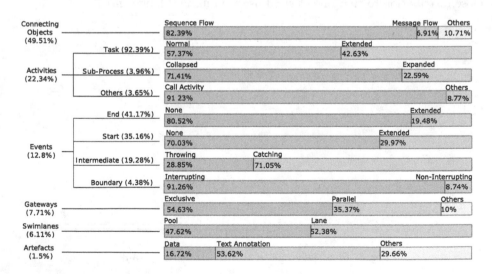

Fig. 4. Use of BPMN elements grouped by categories

4   Complexity of BPMN Models

In this section, we focus on the amount and the types of elements in a model.

4.1   BPMN Models Size

It is well known that the size of a BPMN model (i.e. the number of elements that form the model) affects its understandability [17]. Regarding the 25,590 models that we analysed, we detected that a model presents between 40–41 elements on average, with a median of 31 elements, a standard deviation of 63–64 elements, a maximum number of 3,672 elements and a minimum number of 8 elements. To provide an overview, we classified models based on their size, and we reported those data in Table 2. We divided models into three macro-sets: from 8 to 100, from 101 to 200 and from 201 to over 2000. As we can notice, the first set presents the 96.22% models while the second and third sets respectively present 2.76% and 1.02% of models. The majority of the models (68.61%) are distributed in the classes from 11–20 to 41–50. In particular, the higher concentration is in class 11–20, which presents the 25.5% of models.

Table 2. Number and percentage of models by size

8–100			101–200			201–2000+		
Classes	N° of Models	% of Models	Classes	N° of Models	% of Models	Classes	N° of Models	% of Models
8–10	2 293	8.96	101–110	117	0.46	201–300	149	0.58
11–20	6 615	25.85	111–120	116	0.45	301–400	64	0.25
21–30	3 793	14.82	121–130	115	0.45	401–500	18	0.07
31–40	3 606	14.09	131–140	61	0.24	501–600	8	0.03
41–50	3 543	13.85	141–150	89	0.35	601–700	5	0.02
51–60	1 259	4.92	151–160	52	0.20	701–800	2	0.01
61–70	1 827	7.14	161–170	49	0.19	801–900	3	0.01
71–80	1 128	4.41	171–180	39	0.15	901–1001	2	0.01
81–90	355	1.39	181–190	43	0.17	1001–2000	4	0.01
91–100	202	0.79	191–200	25	0.1	2000+	8	0.03

4.2   Syntactic Complexity of BPMN Models

Studies conducted over other graphical notations (i.e., UML [15,20]) reported that the theoretical complexity of the language (measured by the total number of the modelling elements) is different from the practical complexity given by the number of elements used in the model. To measure the practical complexity of the BPMN modelling notation, we extracted for each model the number of different notation elements used. In Fig. 5 we report the syntactic complexity of BPMN models. As we can see, even considering that the BPMN meta-model describes 85 elements, in our collection of models, the majority (over 54%) uses between 5 and 8 types of elements. Less than 1% of the models use more than 18 different types of BPMN elements. The calculated average of elements types present in a model is 8.85, meaning that a model is designed by using around 8 or 9 types of elements.

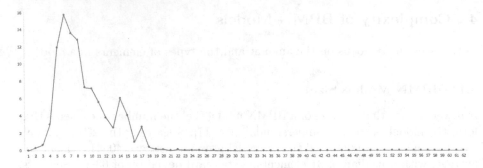

Fig. 5. Syntactic complexity of BPMN models

4.3 Variety of BPMN Subsets

For inspecting the variety of BPMN elements used in models, we used the Hamming Distance. The Hamming Distance refers to the number of different characters in two strings [13]. To calculate the Hamming Distance between two models, we mapped them into binary strings where positive bits signal the presence of a specific BPMN element while negative bits represent their absence. Therefore, for each model, we defined an 85-bit binary string that indicates (using 0s and 1s) which elements each model presents. Then we calculated the Hamming Distance between pair of strings (i.e. between pair of models), and we reported our findings in Fig. 6. As we can notice from the figure, a small percentage (0.66%) of models present a Hamming Distance of zero, which means the same type of BPMN elements forms the models; this highlights a high variability in usage BPMN notation. Most of the models (around 60%) differ from each other for 6–12 BPMN elements. While really few models (0.57%) differ for more than 20 elements, this is also since few models present such an amount of elements. The calculated average dissimilarity between two BPMN models is 8.5, meaning that a model's vocabulary differs from another by 8 or 9 types of elements.

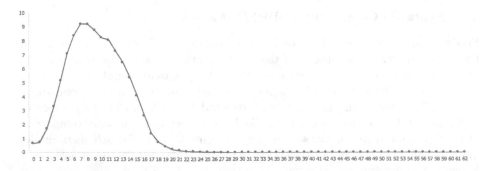

Fig. 6. Hamming distance of BPMN vocabularies

5 BPMN Elements Correlations

To check whether specific BPMN elements are used in combination, we analysed the possible correlation between pairs and groups of elements.

5.1 Correlation Between Pairs of BPMN Elements

To analyse pairs of elements, we determined the covariance matrix of all modelling elements - a square matrix giving the covariance between each pair of elements. Then we adopted the Pearson correlation coefficient (ρ), which corresponds to the covariance between two modelling elements divided by the product of their standard deviations, to normalise the covariance measurement to a value ranging between −1 and 1. Negative values represent inverse correlations, while positive values represent direct correlations. When the value is zero, no correlation is present; while the more the value is close to ±1, the stronger the correlation. In our analysis we distinguished between three degree of correlation: *small* when the value lies below ±0.29, *medium* when the value lies between ±0.30 and ±0.49, *strong* when the value lies between ±0.50 and ±1. For presentation purposes, in Table 3 we report only the most significant pairs of elements that presented a strong direct correlation. None of the elements pairs presented a medium/strong inverse correlation. The pairs presenting a small inverse correlation indicate typed elements at the expense of the not typed ones. It is the case of a model presenting typed elements such as Send Task, Receive Task, Manual Task, Business Task may not present the not typed Task element. Moreover, rarely used elements presented a small inverse correlation with many of the other elements.

For what concerns the *Strong correlations*, we report in the following our interpretation of the obtained results. The highest correlation value corresponds to $\rho = 0.87$, and it is obtained by the pair *Exclusive Gateway-Sequence Flow*;

Table 3. Correlation coefficient between BPMN elements

Element one	Element two	ρ
Exclusive Gateway	Sequence Flow	0.87
Send Task	Receive Task	0.79
Start Event None	End Event None	0.78
Sequence Flow	Task	0.71
Expanded SubProcess	Start None Event	0.67
Inclusive Gateway	Exclusive Gateway	0.65
Exclusive Gateway	Task	0.57
Expanded SubProcess	End None Event	0.56
Pool	Message Flow	0.53
Pool	Lane	0.52

Exclusive Gateways are used to split the workflow into multiple branches using multiple Sequence Flows. It is reasonable, therefore, to notice this strong correlation. *Send Task* and *Receive Task* present a correlation coefficient of $\rho = 0.79$. Those elements are used to model communication aspects regarding a single or multiple communicating processes, and their strong correlation is straightforward. The pair composed by *Start Event None* and *End Event None* have a correlation coefficient of $\rho = 0.78$. Generally, a business process model has at least one start event, and one end event explicitly reported and the not typed once (those named *None*) are the most used. The pair composed by *Sequence Flow* and *Task* presents a strong correlation ($\rho = 0.71$) since tasks are key elements in the design of a business process model, and sequence flows are attached to tasks for defining the control flow. Being and *Expanded Sub-Process* a business process per se, it is also reasonable that its start and end are represented and strong correlation with *Start None Event* and *End None Event*. It results a ρ value of 0.67 and 0.56, respectively. The *Inclusive Gateway* and the *Exclusive Gateway* elements present a correlation coefficient $\rho = 0.65$; this means that the Inclusive Gateway, when used, it is used in combination with other gateways and being the Exclusive Gateway the most used one, the correlation is straight forward. The pair *Exclusive Gateway-Task* presents a correlation coefficient $\rho = 0.57$; this is expected since both elements are among the most used ones. The *Pool* element that is generally used to represent the Owner of a process (e.g. a Company) is often (but not always) used in combination with *Lanes* to distinguish between departments of the same organisation. Therefore a strong correlation with the element Lane ($\rho = 0.52$) has to be expected. The *Pool* element is also used for representing collaborations of processes which generally corresponds to different companies collaborating to achieve some goal. Therefore, it is generally used in combination with the *Message Flow* ($\rho = 0.53$) to represent the exchange of messages between the different parties involved in the process.

5.2   The Combined Use of BPMN Elements

After analysing the correlation coefficient for pairs of BPMN elements, we analysed groups of elements to find those most frequently used in combination.

The Venn diagram in Fig. 7 shows the elements that are most used in combination with the respective percentage of the models in which they are present. We reported only those combinations of elements that appeared together in at least 25% of the models. The dashed lines size are used to group the elements so that the shorter the dash, the higher the percentage of models with that combination of elements. The elements that are primarily used in combination are Tasks and Sequence Flows with a percentage of 96%. Instead, Start Event and the End Event are present in combination in 91% of the models. The above mentioned elements (i.e., Task, Sequence Flow, Start and End Event) form a core set of elements that are mostly used in combination; they are present together in 89% of the models. Additional elements combined with the core set are delimited with a green dashed line. The combination of the core set with Exclusive Gateway is present in 50% of the models, with Pool in 44% of the models, with

Parallel Gateways present in 31% of the models, with Intermediate Events is present in 30% of the models. On the right side of the figure, we reported the combination of Lanes and Message Flows with the Pool element, which respectively are present in combination in 37% and 27% of the models. In particular, these elements are related to the modelling of Organisational aspects (i.e., the owner of the process represented by a Pool of other company departments by Lanes) and the modelling of processes collaborations (i.e., Pools with processes that communicate via message flows). It is worth noticing that typed elements (e.g., User Task, Message Event, Timer Event, etc.) are not included in any of the most popular BPMN element subsets.

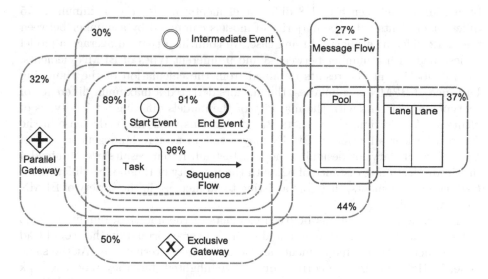

Fig. 7. Most popular BPMN vocabulary subsets

6   Comparison with Muehlen and Recker's Work

Among the literature targeting BPMN, the contribution that most relates to ours is [18]. The authors analysed the usage of BPMN v1.0 over a set of 120 models. The BPMN version that is currently used is BPMN v2.0 which includes new elements with respect to version one. In fact, in our study, we extract data for 85 BPMN elements respect to the 50 elements described in their work. Also, the significant amount of models that we retrieved, 25,590, compared to their 120 models, establishes a stronger base for conducting statistical analysis.

For what concerns the distribution of the occurrence of elements, both the research works highlight a similar trend in the usage of BPMN elements concerning Sequence Flows, Start and End None Events, Tasks, Exclusive Gateway, Pool, Lane, Parallel Gateway and Message Flow being among the most used elements of the notation. The differences in the rankings are evident when we look at those elements that are less used (Multiple Instance, Cancel Events, etc.). However, being among the less used elements, the ranking highly depends on the set of analysed models.

Regarding the frequency distribution of BPMN elements divided by ranks, both the analysis of [18] present a distribution that resembles the power law distribution typical of natural languages (Zipf's law [24]). Thus, we observe that despite the elements added in v2.0 the BPMN language still maintains the frequency distribution of typical natural languages, consisting of a few essential elements and a set of barely used ones.

The set of models we analysed presents a higher variety of elements used, with a maximum of 62 and a higher concentration of models that are composed by a number between 5 and 8 different elements; in [18] the maximum is 15 different elements with the majority of models composed by a number between 6 and 12 different elements. In any case, we confirmed that on average a model is designed with a number between 8 and 9 different types of BPMN elements. In [18], looking at the results obtained from the calculation of the Hamming Distance of BPMN vocabularies, the authors detected a maximum difference of 18 types of elements with the majority of the models differing for 7–8 types of elements. In addition, the authors stated that this value might decrease in the future as wider adoption of BPMN may result in a more homogeneous use of BPMN vocabulary for designing models. Instead, in our results, we obtained a maximum hamming distance of 62, with the majority of the models differing for 6–12 types of elements, therefore assisting to an increase in the variety of BPMN elements used.

Referring to correlations between elements of the notation, our findings do not show any significant (medium and strong) inverse correlation that could lead us to conclude that some elements are actually used alternatively. At the same time, in [18] some inverse correlations are highlighted, but they report a weak interpretation. Instead, for what concerns medium and strong direct correlations, we identified many more correlations to them. However, some correlations they detected did not result from our analysis (e.g., Lane and Message Flow, Start Message and Exclusive Gateway, Start Message and End Terminate).

Referring to the combined usage of BPMN subsets, we obtained results with the same trend of [18]. The pair Task-Sequence Flow it constitutes, for both, the most used combination present in 96% of the analysed models. For all the other combinations we obtained higher percentages denoting a more tight usage of those subsets. Those differences may be due to the different amount of analysed models (i.e., we analysed 25,590 models while they analysed only 120). In addition, compared to [18], we identified more subsets that are used in combination, especially the subsets formed by Intermediate Events and the core set and the subset formed by Message Flows and Pool, that did not emerge from the previous contribution. Finally, in Table 4 we report the comparison of the analyses with Muehlen and Recker's work described above.

Table 4. Comparison of this work with Muehlen's work [18]

	Our work	[18]
Number of BPMN models repositories	6	3
Number of analysed BPMN models	25 590	126
Average complexity of BPMN models	8–9 elements	6–12 elements
Average variety of elements in models	6–12 elements	7–8 elements
Average number of construct's occurrence	✓	✗
BPMN elements distribution over models	✓	✓
BPMN elements frequency distribution	✓	✓
BPMN elements correlations	✓	✓
Combined use of BPMN elements	✓	✓

7 Conclusion, Limitations and Future Work

This research work's objective was to investigate whether the BPMN vocabulary adopted nowadays by model designers shows some particular trends. The main findings confirm that the majority of the models designed with BPMN 2.0 are designed around a core set of elements (i.e., Task, Sequence Flow, Start Event and End Event), resulting in a large portion of the notation that is rarely used. These findings are consistent with those obtained for BPMN 1.0 [18]. These findings are confirmed by comparing the BPMN theoretical complexity, which corresponds to 85, and the practical complexity of the notation, which corresponds to 8–9 elements on average. Our study clearly emphasises the wide usage of the core set of BPMN elements with respect to the most advanced ones. In addition, our results show that some elements are highly correlated, and some are often used in combination; this also can be taken as a reference while preparing training sessions on BPMN. The emerged results can be taken as a reference for guiding the development of BPMN-related tools, which should focus first on providing support for the most used elements and then for the rest of the notation. The results can also affect training activities suggesting a list of elements (the most used ones) that trainers should focus on first before addressing advanced elements (the less used ones).

It should be stated that this research was subject to a limitation given by the analysed models. However, we argue that 25,590 models taken from six different repositories should compose a solid base for generalising the obtained results to general use of BPMN notation.

In the future, we plan to extend our study to incorporate additional models possibly coming from real world applications or additional online repositories. We also want to analyse the usage of the BPMN notation by targeting different application domains to discover possible subsets of BPMN elements that better fit an application domain with respect to another. Finally, we envision a study that involves practitioners and fresh BPMN users to evaluate whether and how the experience, gained by practising BPMN, may lead to a differentiated usage of BPMN elements or whether it is the application domain mostly guides the choice of the BPMN elements to use.

References

1. Rolón, E., Cardoso, J., García, F., Ruiz, F., Piattini, M.: Analysis and validation of control-flow complexity measures with BPMN process models. In: Halpin, T., et al. (eds.) BPMDS/EMMSAD-2009. LNBIP, vol. 29, pp. 58–70. Springer, Heidelberg (2009). https://doi.org/10.1007/978-3-642-01862-6_6
2. Bork, D., Karagiannis, D., Pittl, B.: Systematic analysis and evaluation of visual conceptual modeling language notations. In: Research Challenges in Information Science, pp. 1–11. IEEE (2018)
3. Bork, D., Karagiannis, D., Pittl, B.: A survey of modeling language specification techniques. Inf. Syst, **87**, 101425 (2020)
4. Compagnucci, I., Corradini, F., Fornari, F., Polini, A., Re, B., Tiezzi, F.: Modelling notations for IoT-aware business processes: a systematic literature review. In: Del Río Ortega, A., Leopold, H., Santoro, F.M. (eds.) BPM 2020. LNBIP, vol. 397, pp. 108–121. Springer, Cham (2020). https://doi.org/10.1007/978-3-030-66498-5_9
5. Corradini, F., Ferrari, A., Fornari, F., Gnesi, S., Polini, A., Re, B., Spagnolo, G.O.: A guidelines framework for understandable BPMN models. DKE **113**, 129–154 (2018)
6. Corradini, F., Fornari, F., Polini, A., Re, B., Tiezzi, F.: A formal approach to modeling and verification of business process collaborations. SCP **166**, 35–70 (2018)
7. Corradini, F., Fornari, F., Polini, A., Re, B., Tiezzi, F.: Reprository: a repository platform for sharing business process models. CEUR **2420**, 149–153 (2019)
8. Dijkman, R.M., Dumas, M., Ouyang, C.: Semantics and analysis of business process models in BPMN. Inf. Softw. Technol. **50**(12), 1281–1294 (2008)
9. Fahland, D., et al.: Instantaneous soundness checking of industrial business process models. In: Dayal, U., Eder, J., Koehler, J., Reijers, H.A. (eds.) BPM 2009. LNCS, vol. 5701, pp. 278–293. Springer, Heidelberg (2009). https://doi.org/10.1007/978-3-642-03848-8_19
10. Genon, N., Heymans, P., Amyot, D.: Analysing the cognitive effectiveness of the BPMN 2.0 visual notation. In: Malloy, B., Staab, S., van den Brand, M. (eds.) SLE 2010. LNCS, vol. 6563, pp. 377–396. Springer, Heidelberg (2011). https://doi.org/10.1007/978-3-642-19440-5_25
11. Haisjackl, C., Pinggera, J., Soffer, P., Zugal, S., Lim, S.Y., Weber, B.: Identifying quality issues in BPMN models: an exploratory study. In: Gaaloul, K., Schmidt, R., Nurcan, S., Guerreiro, S., Ma, Q. (eds.) CAISE 2015. LNBIP, vol. 214, pp. 217–230. Springer, Cham (2015). https://doi.org/10.1007/978-3-319-19237-6_14
12. Haisjackl, C., Soffer, P., Lim, S.Y., Weber, B.: How do humans inspect BPMN models: an exploratory study. Softw. Syst. Model. **17**(2), 655–673 (2016). https://doi.org/10.1007/s10270-016-0563-8
13. Hamming, R.: Error detecting and error correcting codes. Bell Syst. Tech. J. **29**, 147–160. Nokia Bell Labs (1950)
14. Heinze, T.S., Stefanko, V., Amme, W.: Mining BPMN processes on Github for tool validation and development. In: Nurcan, S., Reinhartz-Berger, I., Soffer, P., Zdravkovic, J. (eds.) BPMDS/EMMSAD-2020. LNBIP, vol. 387, pp. 193–208. Springer, Cham (2020). https://doi.org/10.1007/978-3-030-49418-6_13

15. Kobryn, C.: UML 2001: a standardization odyssey. Commun. ACM **42**, 29–37 (1999)
16. Kossak, F., et al.: A Rigorous Semantics for BPMN 2.0 Process Diagrams. Springer, Cham (2014). https://doi.org/10.1007/978-3-319-09931-6
17. Mendling, J., Sánchez-González, L., García, F., Rosa, M.L.: Thresholds for error probability measures of business process models. J. Syst. Softw. **85**(5), 1188–1197 (2012)
18. Muehlen, M., Recker, J.: How much language is enough? theoretical and practical use of the business process modeling notation. In: Bellahsène, Z., Léonard, M. (eds.) CAiSE 2008. LNCS, vol. 5074, pp. 465–479. Springer, Heidelberg (2008). https://doi.org/10.1007/978-3-540-69534-9_35
19. OMG: Business Process Model and Notation (BPMN), Version 2.0.2 (2013). https://www.omg.org/spec/BPMN/2.0.2
20. Siau, K., Erickson, J., Lee, L.: Theoretical vs. practical complexity: the case of UML. J. Database Manag. **16**, 40–57 (2005)
21. Wohed, P., van der Aalst, W.M.P., Dumas, M., ter Hofstede, A.H.M., Russell, N.: On the suitability of BPMN for business process modelling. In: Dustdar, S., Fiadeiro, J.L., Sheth, A.P. (eds.) BPM 2006. LNCS, vol. 4102, pp. 161–176. Springer, Heidelberg (2006). https://doi.org/10.1007/11841760_12
22. Wong, P.Y.H., Gibbons, J.: Formalisations and applications of BPMN. Sci. Comput. Program. **76**(8), 633–650 (2011)
23. Zarour, K., Benmerzoug, D., Guermouche, N., Drira, K.: A systematic literature review on BPMN extensions. Bus. Process Manag. **26**(6), 1473–1503 (2020)
24. Zipf, G.K.: On the dynamic structure of concert-programs. J. Abnorm. Psychol. **41**(1), 25–36 (1946)

Capabilities in Crisis: A Case Study Using Enterprise Modeling for Change Analysis

Georgios Koutsopoulos$^{(\boxtimes)}$ (iD)

Department of Computer and Systems Sciences, Stockholm University, Stockholm, Sweden
georgios@dsv.su.se

Abstract. Changing capabilities is a measure that businesses employ as a response to emerging opportunities, threats and necessary adaptations derived from the dynamic environment they operate in. Enterprise Modeling is a discipline that can provide support during the transition of capabilities and facilitate the process. This study is part of a project aiming to develop a method specifically designed for managing capability change using enterprise modeling. This paper's goal is to identify candidate components for the method by exploring semantic consistency among different enterprise models developed in the context of a case study. The reported case study has been conducted in an organization of the public arts and culture sector in Greece that is dealing with multiple difficulties and challenges simultaneously and is driven to adapt its capabilities. Different Enterprise Modeling approaches are employed to capture the wide spectrum of concepts necessary for modeling the complex capability change phenomenon. Potentials for model integration and candidate method components are identified along with business transformation insight derived from the analysis of changes.

Keywords: Capability · Enterprise modeling · Change · Adaptation · Transformation

1 Introduction

Organizations have always been dealing with dynamic environments, however, the modern conditions, taking into consideration the digital transformations, have led to a degree of dynamism that constantly raises changes, opportunities and threats for active businesses [1]. Change, as a phenomenon, is not an exception anymore. It is gradually becoming the standard for businesses and since change and strategy are two concepts inextricably linked, change drives strategy and strategy drives change [2]. Strategy always concerns planning, deciding and acting in a way that will result in fulfilling business goals [3]. Organizations aim strategically for flexibility and resilience as a way to improve their efficiency towards dealing with change.

Analyzing change is an essential part of strategy, and decision-making bodies often try to anticipate changes and plan measures to tackle them in advance. However, this is not always possible, especially when the changes are the result of a crisis that affects the organization in several unpredictable aspects, as for example during the COVID-19

© Springer Nature Switzerland AG 2021
R. A. Buchmann et al. (Eds.): BIR 2021, LNBIP 430, pp. 100–114, 2021.
https://doi.org/10.1007/978-3-030-87205-2_7

crisis. Nevertheless, a popular view from the crisis management literature suggests that there is a thin line between crisis and opportunity [4]. In this case, the outcome depends on the organization's response, and, as a result, changing as a response to a crisis requires support through analyses from multiple perspectives.

A variety of business analysis methods can assist the analysis of change. The approach examined in this paper is Enterprise Modeling (EM). By using multiple enterprise models, it is possible to study several aspects of a change. In this study, it is examined how EM can be employed as a way to facilitate the analyses and provide support via the structuring of information. The specific EM approaches used are capability [5], goal [6] and value modeling [7] which set capability, goal and value respectively as the focal modeling point.

This paper is part of an ongoing Design Science Research (DSR) [8, 9] project aiming to provide methodological and tool support for the management of changing organizational capabilities. To date, a capability change meta-model has been presented, aiming to serve as the basic component for a modeling method. The objective of this paper is to *identify other modeling approaches as candidate components for the method* by examining the *semantic consistency between elements of different models* along with potentials for model integration and integration points. This is achieved by reporting on a case study where the meta-model has been applied in conjunction with other modeling methods.

The reported case study has been performed in Veria Art Center, a public organization in the arts and culture sector in the municipality of Veria, Greece. The study focuses on the organization's capability to organize festivals relevant to arts and culture. While adapting to the long-term ongoing effects of the Greek government-debt crisis and the reduction of resources, the organization needs to face the COVID-19 crisis in parallel, with the social distancing regulations disrupting the vast majority of planned activities and enforcing alternative ways to deliver equivalent value to the municipality's residents. The objective of the case study was to collect data about the organization's changing capabilities, support and facilitate the transition via the creation of enterprise models that will support the organization's decision-making processes.

The rest of the paper is structured as follows. Section 2 briefly provides the background and research related to this study. Section 3 describes the research strategy and method selected for this study. Section 4 presents the studied organization and its capabilities, along with the specific change cases that have been explored. Section 5 describes the results and models developed during the case study along with identified potentials for model integration. Section 6 discusses the results and lessons learned from the study. Finally, Sect. 7 provides concluding remarks.

2   Background and Related Literature

The main topics relevant to this study are the concept of capability, the discipline of EM, and organizational change and crises.

Organizational change has been widely researched under various terms like change, transformation or adaptation used either interchangeably or expressing differences in scopes [10]. It is commonly driven by (i) rational adaptation to changing environmental conditions, (ii) strategic decision-making in the forms of plans and choices or (iii) both options combined with organizational inertia. In this way the existing organizational change theories are classified as deterministic, voluntaristic and reconciling respectively. A commonly neglected factor is the causality of change whose implementation should be required in methods that tackle the complex nature of ongoing changes [11].

This complexity is significantly increased when dealing with crises. A crisis has been defined as a threat to the performance and sustainability of the organization [12], being a time or state of affairs without stability which involves an impending decisive change. However, many researchers consider that a crisis' outcome can be highly desirable or highly undesirable [13] since it is both a danger and an opportunity [14] and its destructive side is a required condition for an organization's development [15].

EM approaches' purpose is to capture, elicit, document, deliberate and communicate the abovementioned complexity. This is achieved by capturing various organizational aspects that are relevant for the given modeling purpose, for example, processes, goals, business rules and concepts [6]. What bears great significance is the integration of these aspects in a single view. Thus, an enterprise model often consists of integrated sub-models, each focusing on a specific relevant aspect. The term integration refers to consistency between a model's constituent sub-models. It requires co-existence of concepts between the sub-models. These concepts are used as integration points.

Capability describes one organizational aspect. Within this project, it is defined as a set of resources and behaviors, with a configuration that bears the ability and capacity to enable the potential to create value by fulfilling a goal within a context [16]. It is associated to essential core business concepts like actor, goal, process, resource and context [17, 18]. Apart from encompassing all these concepts, the value of capability lies in the fact that it is considered the missing link in business/IT transformation, serving as the baseline for change management, strategic planning and impact analysis [19].

Several types of modeling approaches have included the concept of capability. There are stand-alone approaches like Capability-Driven Development (CDD) [17], Enterprise Architecture Frameworks like NATO Architecture Framework (NAF) [20] or new notations like CODEK [21]. For a detailed review of the literature, an extensive list of the capability approaches and meta-models has been presented in a previous step of this project [22]. Many of the included approaches include case studies similar to this one.

3   Methodology

In this design cycle of DSR the aim is to fine-tune the existing requirements and test the meta-model's consistency with existing methods. This concerns an iterative movement between eliciting requirements for the artifact and designing and developing the artifact [8]. Case study is appropriate for in-depth investigation of a contemporary phenomenon which exists in real-life contexts and, in addition, the researcher has no control over the studied events [23]. The main goal of the case study was to provide analysis and support for change planning regarding the organization's capabilities and explore semantic consistency in the developed models in order to identify potential integration points. Sub-goals include (i) understanding the business, (ii) identifying the main capabilities, (iii) exploring how Veria Art Center deals with changes and (iv) studying how a specific capability changes and supporting planning for its desired state.

The selected data collection method is guided face-to-face interview. Guided interviews are a type of unstructured interviews, during which the researcher only prepares a few questions in order to guide the conversation but let the participant "tell his/her story" [24]. The questions used to guide the discussion were directly derived from the sub-goals, for example "How does your organization deal with changes?".

The sampling strategy was convenience and purposive sampling [25] to ensure that the required information is collected. In particular, all the managers and heads of the organization's departments have participated and provided insight on the researched phenomenon from their own perspective. As a result, one group interview and six individual interviews were conducted, with an average duration of about one hour each.

After the interviews were completed, the data was organized and used for the development of enterprise models focusing on different aspects related to the studied capability. The models were developed in two two-hour long participatory modeling sessions [26], where the author facilitated the process with the participants as domain-experts. It was examined how various types of enterprise models included the necessary information types to analyze the change phenomenon. Apart from the integrated version (Fig. 1) of the capability change meta-model [5] and its extension [27] using the Unified Modeling Language (UML) [28], the selected modeling methods were e3value [7] to document value outcomes and 4EM (For Enterprise Modeling) [6] to document goals. The selection of modeling methods was based on the participants' opinions on what they needed support with, for example, they are already using structured processes, so process models were deemed as least useful.

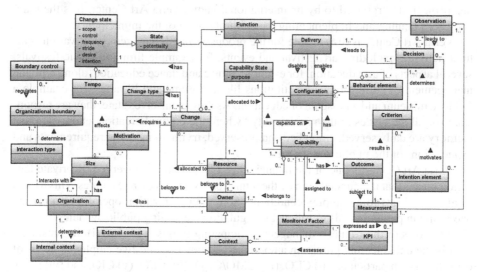

Fig. 1. The integrated capability change meta-model.

The purpose of using the meta-model was to capture changes in the business capability that was studied. The meta-model is based on a framework [22] which includes the functions of change, i.e. (i) observation, (ii) decision and (iii) delivery. Observation is captured by including internal and external context elements, which consist of monitored factors which are expressed as KPIs. According to the meta-model, a capability has one or more configurations that lead to the realization of the capability. A configuration has resources allocated to it and also consists of behavior elements like activities, processes and services. The realization of the capability produces one or more outcomes that fulfill the initial intention elements that justify the existence of the capability. Deciding to change depends on these intention elements and other criteria derived from contextual factors. Delivery concerns the transition from one configuration to another. Furthermore, the meta-model includes ownership elements to capture who owns the capability, the resources and the change project. In case different owners exist, the meta-model includes elements to describe their interaction and boundaries. Finally, the states of the capability and change are included along with their dimensions, as identified in [29]. The meta-model includes both abstract and specific concepts.

4   The Case Study

The organization where the case study has been conducted is located in Veria, a small urban city in northern Greece. The organization's official Greek name means Public Benefit Enterprise for Multiple Development of the Municipality of Veria, however, in this project, it is referred to by its international name, Veria Art Center. It is the main body for planning and implementing the cultural policy of the municipality.

Veria Art Center is responsible for the production and management of various cultural activities including art festival and events, for example film and music festivals, several aspects of art education, for example, music and dance education, along with the management of culture-related institutions like museums and libraries, for example a folklore museum and a museum of education. As they themselves state, they are responsible for all the places in the municipality "where the cultural creation, expression and memory are daily served, embodied and preserved, trying to meet the requirements and expectations of the public and the city".

Veria Art Center is one of the municipality's legal entities governed by private law, which means that it is funded both by the municipality and its own resources. This hybrid legal status leads to a complex condition where the organization operates as a private one while in parallel needs to follow the regulations regarding public agencies.

The roles of the Center's managers participating in this study are (i) the president, who is the only electing politician among the participants, and five employed heads of departments, in particular, (ii) CEO, (iii) CMO/COO, (iv) CFO, (v) CIO, and (vi) CTO.

4.1   Case Overview

The main capabilities of Veria Art Center have been identified to be the production and organization of art festivals, the provision of art education and the management of culture-related institutions. All four capabilities have been affected by the conditions that affect the whole organization, however, the study focused on the *organization of art festivals* because it was the capability whose adaptation interested mostly the Center.

The capability has been subjected to a variety of minor changes over the years, however, in this study, the point of focus is on two major changes that are currently affecting it and the planning for a future change. Regarding the dimensions of a capability which have been identified in [29], a capability's scope is to fulfill a goal in comparison to avoiding problems. It actively produces value and is owned by a single organization, Veria Art Center.

In its normal configuration, the capability is a combination of producing and organizing festivals, which means that there is a continuous collaboration among Veria Art Center and external producers of art and culture events. The resources required for delivering value are the existence of appropriate infrastructures, in particular, the three stages that the organization owns and manages, which means that these also require maintenance staff. Additionally, the stages require technical staff operating the required equipment that is essential for the realization of a festival, for example light and sound equipment operators. The knowledge and expertise required for this operation is another important resource that cannot be omitted. There are also the financial resources required, not only for the abovementioned resources' cost, but also for promoting the events via digital and physical campaigns. Finally, a special type of resource is reputation, since the external collaborations are built on it and it is also required for a successful promotional campaign. An identified weak spot in the current configuration of the capability is the lack of a method to collect data that would allow evaluating the success of an event, since everything is performed physically, the majority of events is outdoors and the attendance is free.

The following sections will describe the changes, explaining the causes of change, the modification of the capability's configuration according to the functions introduced in the change framework [22] and the dimensions of change [29] for each case. A noteworthy fact is that despite the successful handling of change, the organization does not appreciate its current state. As one of the participants stated, "An organization should not rely on outside the box solutions".

Change 1: Transition to Lean

Observation: The initial change applied to the art festival organization capability was forcefully imposed due to the Greek government-debt crisis and its ongoing long-term effects, especially the ones impacting the public sector. The availability of resources was significantly limited due to consecutive budget cuts. In practice this was imposed as an updated legislation that did not require dismissing existing employees but mitigated the public organizations' ability to employ replacements for any retiring human resources. In addition, a restriction regarding expert employees has been applied, which means that replacement employments usually result in a downgraded level of expertise.

Decision: Transition from normal to lean.

Delivery: The major effect of this change is the gradual reduction of available human resources, both in quantity and quality. In terms of quantity, the point where there were not enough people to perform necessary tasks was reached. However, it was tackled by adopting a flexible work schedule after a negotiation process between the managers and the employees. This enabled a level of flexibility that allowed optimal use of human resources and the actual result was an increase in the number of delivered events despite the reduction in human resources. In terms of quality, a lack of several pieces of expertise was identified after the retirement of specific employees and, considering the legislative restrictions, the problem was solved by either spending financial resources for training remaining employees or contracting external collaborators. This depended on the frequency of events that required a specific type of expertise. For example, on one hand, knowledge of operating a specific sound console that was required twice a year was fulfilled by external contracting, on the other hand, operating the lights console on Veria Art Center's own stage required training of an existing employee, since it is required several times monthly.

State: Active, *Tempo:* Slow, *Frequency:* Continuous, *Desire:* Undesirable, *Stride:* Incremental, *Scope:* Adaptation, *Control:* Emergent, *Intention:* Unintended.

Change 2: Transition to Digital

Observation: While the effects of transitioning to a lean configuration of the capability are still influencing the organization, another emergent change became necessary due to the COVID-19 pandemic crisis. Traditionally, the art events and festivals have been held exclusively with a physical presence required, both for the performers and the audience, a condition which has been forbidden due to the social distancing restrictions applied in Greece as a response to the pandemic. As a result, Veria Art Center had to adapt to the emergency and invent new variations of the capability that serve the same goals and produce the same value while complying with the emergent regulations.

Decision: Transition from lean to digital.

Delivery: The easiest way to achieve this task is the delivery of digital online events and festivals. Before the change was applied, it was unanimously agreed within the organization that the digital events cannot deliver the same value as the physical ones, therefore the replacement of their events by their online versions is a temporary solution. An advantage of the digital configuration of the festival organization capability is the significant reduction of required resources. There is no need for physical infrastructures and their maintenance. In addition, less equipment and operators are required. The promotion of a digital event is also significantly cheaper since there is no need for physical means like flyers and posters in the city. The contracting process is also different since the production costs are sometimes cheaper for a home-based event for external collaborating artists and producers. Another identified opportunity is the collection of accurate data, a condition which allows for evaluating the success of an event. Finally, the most important resource becomes the platform that hosts not only the online event but also its promotion. In this case, the social media have provided an ideal solution with a minimum cost. One interesting fact is that the digital events attracted a significantly higher number of audience, since the physical restrictions of a local event are lifted in a digital platform. This resulted also in increased reputation, since the audience in a digital event joins nationwide.

State: Active, *Tempo:* Fast, *Frequency:* Once, *Desire:* Undesirable, *Stride:* Revolutionary, *Scope:* Transformation, *Control:* Emergent, *Intention:* Unintended.

Change 3: Transition to New Normal

Observation: The two emergent and undesired changes led to two different configurations of the capability. Both the lean and digital configurations of the capability have advantages and disadvantages, and the goal of Veria Art Center is to preserve as many advantages as possible from the two configurations and utilize them when the COVID-19 crisis restrictions are over. Regarding the lean configuration, as long as the delivery of the capability is viable with a given set of resources, there is no need to return to a less efficient configuration.

Decision: Transition from digital to a new normal, combining lean and digital.

Delivery: The ability to predict the effect of a change in a planning phase is limited, but the scope can be clear. The digital configuration has provided an expanded audience that should not be ignored after the pandemic crisis. One potential hybrid configuration of the capability is to have parallel physical and digital events using streaming services. The digital configuration's data collection potential also has value that should be used. Initially, event success criteria can be established. In addition, using the data collected from the digital participation will provide a decision criterion for evaluating whether it is worth to expand the physical event's radius, so that a greater area is considered "local" and is its residents are directly invited by expanding the marketing campaigns of the event. A hybrid physical and digital marketing campaign has a reduced cost compared to an exclusively physical one and the gain can potentially outweigh the cost. All the digital attributes are planned to be implemented in a mobile app that will handle all the required functionalities both for physical and digital events, for example, reservations, tickets, ratings and schedule. Finally, all the lean attributes are planned to remain unless there are changes in the legal context that allow for more flexibility.

State: Inactive, *Tempo:* Slow, *Frequency:* Once, *Desire:* Desirable, *Stride:* Incremental, *Scope:* Adaptation, *Control:* Planned, *Intention:* Intended.

5   Analysis of the Case

This section presents the models developed by structuring the information of the Veria Art Center case in order to address the study's goal and sub-goals.

5.1   Understanding the Business: Value Transactions

In order to better understand the business in terms of delivered capabilities the organization preferred a depiction of the value delivered to the municipality. For this reason, an e3value model was created that captures all the main value activities of Veria Art Center along with all the external collaborators.

In particular, the model includes the capabilities to manage/produce an art festival, to manage art and culture institutions and provide art and culture education to the residents of the municipality of Veria, as shown in Fig. 2. The behavioral aspect of the capabilities has been modeled as value activities according to the e3value notation.

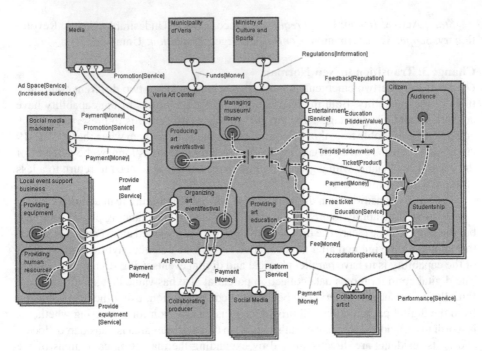

Fig. 2. The e3value model for the Veria Art Center case.

The other actors and market segments in the model are the citizens that receive value both as audience and students by the delivery of capabilities provided by the organization, the municipality of Veria which funds the capabilities, the Ministry of Culture and Sports that determines the regulations, the media, social media and specialized marketers that provide their services for a fee, collaborating producers and independent artists that participate in the festivals and local event support businesses that rent specific missing resources, in particular, equipment and staff. An interesting finding lies in the fact that there are hidden values in the delivery of the art festival capability, a point that was heavily emphasized by all the organization's participants, both in the individual and group meetings. For example, a music festival does not simply provide entertainment, it also has educational value but also creates cultural trends on a local level.

5.2 Desired State of the Capability

The participants considered modeling the goals of the "new normal" configuration essential to document, therefore a 4EM Goals model (Fig. 3) was developed to decompose and analyze the requirements regarding the future of the capability.

The model depicts goal G1, to deliver a successful festival, as the main goal, even if there are higher level goals which reflect on both the visible and hidden values delivered by the realization of the capability (G4–6, G34). The goal is initially decomposed into the requirements for organizing a festival (G3) and making it successful (G2), a fact that motivates the establishment of success criteria, based on several aspects related

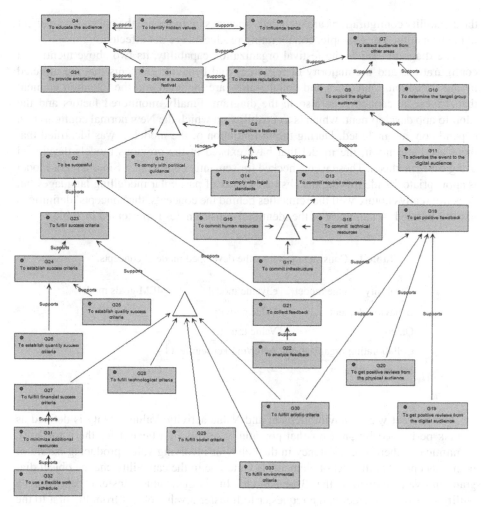

Fig. 3. A 4EM Goals model depicting the desired state of the capability.

to relevant context factors and this is reflected in goals G23–33. G7–11 reflect on the organization's desire to benefit from the expanded audience gained by using the digital configuration of the festival organization capability. G12 and 14 concern contextual requirements while G13 and G15–17 depict resource allocation goals. Finally, G18–22 depict the introduction of data collection from the audience for event evaluation.

5.3 Capability Transition and Integration Points

The transition of capabilities has been modeled by instantiating a fragment of the capability change meta-model. To reduce model complexity and size, the object diagram only includes one of the three presented transitions, the transition from the digital configuration to the desired new normal. In addition, for the same reasons, not all components of the capability configurations are included. The ones included are the ones with an effect in the transition. For example, collaborating producers are not affected.

The diagram depicts the festival organization capability, its two above mentioned configurations and the majority of resources and behavior elements that are affected by the transition. Outcomes and collaborators are included and the intention element that drives the capability is also in the diagram. Finally, monitored factors and the Mobile app development, which is a capability on which the New normal configuration depends on, are included. During the instantiation of the model, it was identified that there are elements in the model that have existed in the previous models developed using other methods. These were checked for semantic consistency. Since a meta-model is appropriate for identifying the abstract syntax of particular modelling languages but does not really capture well the semantics behind the concepts, the concepts' definitions have to be used. A summary of the identified consistencies is presented in Table 1.

Table 1. Consistency among the developed models' concepts.

Capability change model	e3value model	4EM goals model
Behavior element	Value activity	–
Outcome	Value transfer	–
Collaborating organization	Actor/market segment	–
Intention element	–	Goal

The first set was Behavior element and Value activity. Value activity is defined as "a task performed by an actor that potentially results in a benefit for the actor." [7]. Semantically, there is consistency in the behavior-involving value-producing nature of both concepts. Another set of elements is Outcome in the capability change object diagram and Value transfer in the e3value model. In e3value, value transfer is defined as "a willingness of a provider and a requester to transfer a value object from the first to the second" [7]. The Outcome element also concerns the value produced by the realization of the capability, a fact that indicates consistency with the Value transfer concept. The combined objects of an Organization and its Interaction type in the capability change diagram are consistent with the Actor and Market segment elements, which are generally defined as independent entities, which are responsible for their own well-being [7], taking into consideration that there are value transfers in the form of resources exchanged. Finally, the abstract Intention element in the object diagram is checked for consistency with the Goal element in the 4EM Goals model. A Goal in 4EM is defined as "a desired state of the enterprise that is to be attained" [6] and the Intention element is an abstract meta-element aiming to include all the concepts referring to intentions like goals, objectives and desires. As a result, the concepts are semantically consistent.

These have been marked in Fig. 4 as possible integration points for a future integrated depiction of the case study phenomenon. In particular, the points highlighted with red are integration points with the e3value model, and the point highlighted with green is a possible integration point with the 4EM Goals model.

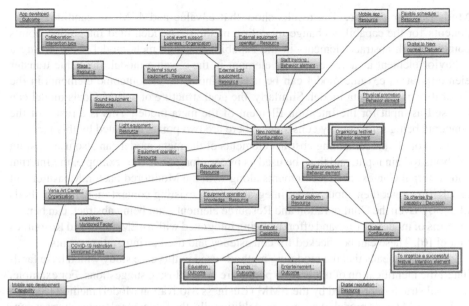

Fig. 4. The object diagram depicts the transition from the digital configuration to the new normal configuration and the highlighted integration points with e3value (red) and 4EM Goals (green). (Color figure online)

6 Discussion

Despite the fact that the case study was conducted within the COVID-19 pandemic, there were no limitations to the application of the methods, in terms of time or effort. The only limitation was the participants lack of familiarity with the modeling methods but it was overcome with commitment to the project. So, the result is that the case study provided valuable insight both from a business perspective and a modeling perspective.

Business Perspective. Initially, the results of the study provide additional support to the crisis management views that associate the concepts of crisis and opportunity. The analysis emphasized on the unexpected benefits that emerged during the adaptation, for example, the expanded audience that can be used to improve the capability when the conditions allow the transition to the New normal configuration. The organization's survival is threatened by the existing crisis legislation, therefore, new innovative capability configurations will be necessary in the near future. Another noteworthy fact is that during the analysis, the importance of hidden values was emphasized, a concept which has been neglected in the majority of modeling methods.

In the bigger picture, the results point out the importance and benefits derived from structuring and analyzing the organization's data using EM. The task would have been beneficial even before the crises, thus, within the current context, the analysis is crucial since "it enables knowing your limits", as stated by a participant. Overall, it seems safe to assume that during a crisis, the value of a multi-perspective analysis method like EM is augmented, since it can be the determining factor between a desirable or undesirable outcome of a crisis.

Modeling Perspective. The results show that e3value and 4EM are candidate components for the capability change modeling method. The identified integration points coincide with abstract elements in the capability change meta-model. Considering a behavior element as a value activity element in the e3value model, the value transfer elements in an e3value model can be used as input for the outcome elements in the capability change meta-model. Similarly, the goal structure of a 4EM Goals model can be used as input for the Intention Element of the meta-model. This is a result of the concepts being semantically consistent. There are two points indicated by this fact.

First of all, the remaining abstract elements of the meta-model can also use existing methods to gain input, and thus, be used as integration points. For example, the Internal and External Context abstract elements along with the Monitored Factor, Measurement and KPI can get their input via an integration with existing context modeling methods like [30]. Even the Configuration and Resource elements, not being abstract, can benefit in terms of modeling time and effort by being integrated with a 4EM Actor and Resources model [6]. These can be checked for consistency and used as integration points.

This also means that the modeling methods used in this case study which have facilitated the identification of integration points are not exclusive suggestions. For example, the goal structure provided by the BMM (Business Motivation Model) method [31] can replace 4EM Goals model, if necessary. Additionally, the Behavior element abstract concept can be used as an integration point with any process or service model that provides decomposition, using languages like BPMN [32] or UML Activity diagrams [28]. Furthermore, the required value structure existing in e3value could be replaced by VDML's (Value Delivery Modeling Language) [33] value propositions.

7   Conclusions

This paper reported on a case study conducted in Veria Art Center, a public organization in the arts and culture sector in the municipality of Veria, Greece. The study focused on the capability to organize festivals relevant to arts and culture. Facing a plethora of challenges, the organization required multi-perspective analyses, a fact which enabled the application of multiple modeling perspectives. These were performed in the form of several enterprise models, in particular, value, goal and capability change models. The models were used not only to facilitate and guide the transition of the organization's capability, but also to identify consistency of concepts and integration points using the capability change model as the focal point of the exploration. Semantic consistency between existing methods' elements and the abstract elements of the capability change meta-model has been identified. This study has successfully met its objective and contributes towards the development of a methodology designed for capturing capability change by identifying e3value and 4EM as candidate components and indicating potentials for model integration. Presenting an integration of the different modeling approaches is a potential topic for future research. Additionally, the findings are applicable to any other context that is facing capability change and resource reallocation challenges, since the methods applied use concepts and modeling elements that are generic and not restricted by the given domain and context and there has been no need to adapt any aspects of the methods to fit the case study.

Acknowledgments. The author would like to express gratitude to all the Veria Art Center participants for their involvement in the case study presented in this paper.

References

1. van Gils, B., Proper, H.A.: Enterprise modelling in the age of digital transformation. In: Buchmann, R.A., Karagiannis, D., Kirikova, M. (eds.) PoEM 2018. LNBIP, vol. 335, pp. 257–273. Springer, Cham (2018). https://doi.org/10.1007/978-3-030-02302-7_16
2. Hoverstadt, P., Loh, L.: Patterns of Strategy. Routledge/Taylor & Francis Group, London/New York (2017)
3. Cunliffe, A.L.: Organization Theory. SAGE, Los Angeles (2008)
4. Alas, R., Gao, J.: The connections between crisis and enterprise life-cycle stages. In: Alas, R., Gao, J. (eds.) Crisis Management in Chinese Organizations, pp. 63–73. Palgrave Macmillan, London (2012). https://doi.org/10.1057/9780230363168_6
5. Koutsopoulos, G., Henkel, M., Stirna, J.: Conceptualizing capability change. In: Nurcan, S., Reinhartz-Berger, I., Soffer, P., Zdravkovic, J. (eds.) BPMDS/EMMSAD-2020. LNBIP, vol. 387, pp. 269–283. Springer, Cham (2020). https://doi.org/10.1007/978-3-030-49418-6_18
6. Sandkuhl, K., Stirna, J., Persson, A., Wißotzki, M.: Enterprise Modeling: Tackling Business Challenges with the 4EM Method. Springer, Heidelberg (2014). https://doi.org/10.1007/978-3-662-43725-4
7. Gordijn, J., Akkermans, H.: Value webs: understanding e-business innovation. The Value Engineers B.V. (2018)
8. Johannesson, P., Perjons, E.: An Introduction to Design Science. Springer , Cham (2014). https://doi.org/10.1007/978-3-319-10632-8
9. Hevner, A., March, S.T., Park, J., Ram, S.: Design science in information systems research. MIS Q. **28**, 75–105 (2004). https://doi.org/10.2307/25148625
10. Maes, G., Van Hootegem, G.: Toward a dynamic description of the attributes of organizational change. In: (Rami) Shani, A.B., Woodman, R.W., Pasmore, W.A. (eds.) Research in Organizational Change and Development, pp. 191–231. Emerald Group Publishing (2011)
11. Zimmermann, N.: Dynamics of Drivers of Organizational Change. Gabler, Wiesbaden (2011). https://doi.org/10.1007/978-3-8349-6811-1
12. Hutchins, H.M., Wang, J.: Organizational crisis management and human resource development: a review of the literature and implications to HRD research and practice. Adv. Dev. Hum. Resour. **10**, 310–330 (2008)
13. Keeffe, M.J., Darling, J.R.: Transformational crisis management in organization development: the case of talent loss at Microsoft. Organ. Dev. J. **26**, 43–58 (2008)
14. Ulmer, R.R., Sellnow, T.L., Seeger, M.W.: Effective Crisis Communication: Moving from Crisis to Opportunity. SAGE Publications, Thousand Oaks (2018)
15. Pauchant, T.C., Mitroff, I.I.: Transforming the Crisis-Prone Organization: Preventing Individual, Organizational, and Environmental Tragedies. Jossey-Bass Publishers, San Francisco (1992)
16. Koutsopoulos, G.: Managing Capability Change in Organizations: Foundations for a Modeling Approach (2020). http://urn.kb.se/resolve?urn=urn:nbn:se:su:diva-185231
17. Sandkuhl, K., Stirna, J. (eds.): Capability Management in Digital Enterprises. Springer , Cham (2018). https://doi.org/10.1007/978-3-319-90424-5
18. Wißotzki, M.: Exploring the nature of capability research. In: El-Sheikh, E., Zimmermann, A., Jain, L.C. (eds.) Emerging Trends in the Evolution of Service-Oriented and Enterprise Architectures. ISRL, vol. 111, pp. 179–200. Springer, Cham (2016). https://doi.org/10.1007/978-3-319-40564-3_10

19. Ulrich, W., Rosen, M.: The business capability map: the "rosetta stone" of business/IT alignment. Cut. Consort. Enterp. Archit. **14**, 1–23 (2011)
20. NATO: NATO Architecture Framework v.4 (2018). https://www.nato.int/nato_static_fl2014/assets/pdf/pdf_2018_08/20180801_180801-ac322-d_2018_0002_naf_final.pdf
21. Loucopoulos, P., Kavakli, E.: Capability Oriented Enterprise Knowledge Modeling: The CODEK Approach. In: Karagiannis, D., Mayr, H., Mylopoulos, J. (eds.) Domain-Specific Conceptual Modeling, pp. 197–215. Springer, Cham (2016). https://doi.org/10.1007/978-3-319-39417-6_9
22. Koutsopoulos, G., Henkel, M., Stirna, J.: An analysis of capability meta-models for expressing dynamic business transformation. Softw. Syst. Model. **20**(1), 147–174 (2020). https://doi.org/10.1007/s10270-020-00843-0
23. Schwandt, T.: The SAGE Dictionary of Qualitative Inquiry. SAGE Publications, Inc., Thousand Oaks (2007)
24. Gubrium, J., Holstein, J., Marvasti, A., McKinney, K.: The SAGE Handbook of Interview Research: The Complexity of the Craft. SAGE Publications Inc., Thousand Oaks (2012). https://doi.org/10.4135/9781452218403
25. Denscombe, M.: The Good Research Guide: for Small-Scale Social Research Projects. McGraw-Hill/Open University Press, Maidenhead (2011)
26. Stirna, J., Persson, A.: Enterprise Modeling: Facilitating the Process and the People. Springer, Cham (2018). https://doi.org/10.1007/978-3-319-94857-7
27. Koutsopoulos, G., Henkel, M., Stirna, J.: Improvements on capability modeling by implementing expert knowledge about organizational change. In: Grabis, J., Bork, D. (eds.) PoEM 2020. LNBIP, vol. 400, pp. 171–185. Springer, Cham (2020). https://doi.org/10.1007/978-3-030-63479-7_12
28. Object Management Group (OMG): OMG® Unified Modeling Language® (2017). https://www.omg.org/spec/UML/2.5.1/PDF
29. Koutsopoulos, G., Henkel, M., Stirna, J.: Modeling the dichotomies of organizational change: a state-based capability typology. In: Feltus, C., Johannesson, P., Proper, H.A. (eds.) Proceedings of PoEM 2019 Forum, pp. 26–39. CEUR-WS.org (2020)
30. Koç, H., Sandkuhl, K.: Context modelling in capability management. In: Sandkuhl, K., Stirna, J. (eds.) Capability Management in Digital Enterprises, pp. 117–138. Springer, Cham (2018). https://doi.org/10.1007/978-3-319-90424-5_7
31. Object Management Group (OMG): Business Motivation Model Version 1.3 (2015). https://www.omg.org/spec/BMM/1.3/PDF
32. Object Management Group (OMG): Business Process Model and Notation (2011)
33. Object Management Group (OMG): Value Delivery Modeling Language v.1.1 (2018). https://www.omg.org/spec/VDML/1.1

Post-merger Integration Specific Requirements Engineering Model

Ksenija Lace$^{(\boxtimes)}$ and Marite Kirikova

Riga Technical University, Riga, Latvia
marite.kirikova@rtu.lv

Abstract. A goal of post-merger integration (PMI) is to combine several companies into one common entity so that Mergers and Acquisitions (M&A) strategic objectives can be achieved (some examples of objectives are growth, optimization, strength, etc.). As an integration process is related to large amounts of information to be acquired, processed, and applied during integration planning and execution phases, knowledge management can be perceived as one of the key supporting factors during the PMI. Additionally, although a PMI should cover integration of all system perspectives – social, cyber and physical; today, in many organizations, the main impact is on the cyber perspective, as it forms the infrastructure for the functioning of the whole system. This leads us to the assumption that the PMI process should be organized in such a way that information system integration becomes its central part including knowledge management activities. This research explores a requirements engineering model that can be incorporated into the PMI phase of any M&A initiative. In comparison with common requirements engineering approaches, this model extends them and is adjusted to support the following PMI features: (1) Focus on the integration and M&A strategic goals. (2) Extensive knowledge management with accumulating M&A experience, (3) Coverage of both explicit and tacit knowledge in all three system perspectives – social, cyber and physical.

Keywords: Mergers & Acquisitions · Knowledge management · Requirements engineering

1 Introduction

The post-merger integration (PMI) phase of a Mergers and Acquisitions (M&A) initiative is the constituent where the company integration is executed, i.e., during this phase several companies are combined into one entity, so that stated M&A strategic goals are achieved [27, 34]. PMI usually involves an initial phase, that is an investigation of the internal capabilities of the companies involved, a due diligence phase of current enterprise architectures, a decision phase and creation of an integration plan, and finally an integration execution phase [27, 28]. PMI is concentrated on the consolidation of different organizational aspects – social, cyber, and physical. However, the cyber perspective, or the integration of information systems (IS), can be perceived as the enabling part of

© Springer Nature Switzerland AG 2021
R. A. Buchmann et al. (Eds.): BIR 2021, LNBIP 430, pp. 115–129, 2021.
https://doi.org/10.1007/978-3-030-87205-2_8

an integration, as it forms the basis and support for integration and future functioning from the points of view of the other two perspectives [28, 33]. At the same time, PMI is usually associated with high levels of complexity and challenges. All this leads to the conclusion that, in order to minimize integration failure risks, PMI should be organized around IS integration [30, 31].

As requirements engineering (RE) is essential in IS development in general, it is logical to use it as a basis for the PMI IS integration process. However, the main goal of common RE is, usually, a new system development, not an existing system integration. Existing system integration can require some adjustments in a RE process [24–26, 29]. Additionally, PMI is usually related to the need to acquire, process and share large amounts of information in a rather short period of time, and is accompanied by risks related to human factors and security limitations [13, 17, 18, 20, 21]. Thus, in the PMI case, RE requires efficient knowledge management to support the process in which individual knowledge is transformed into organizational knowledge that, later, can be accessed and reused.

Within the scope of this research, the following research methodology was used – existing research on the PMI process was explored and evaluated from the perspective of its alignment with requirements engineering approaches. The outcome was that PMI specific RE approach quality criteria were identified. Research on existing RE models was conducted from the perspective of how well they can support identified PMI specific quality criteria and, as an outcome, the required adjustments in the current RE approaches were identified. Based on the required adjustments, the new PMI specific RE model was proposed and its applicability was tested with a real-life PMI initiative example.

The paper is organized as follows: in Sect. 2, PMI specific RE aspects are discussed and quality criteria for PMI specific RE model are defined; in Sect. 3, existing RE approaches, which are related to some PMI RE aspects, are selected and evaluated using stated quality criteria (as a summary, already existing reusable RE model parts are identified and required adjustments for PMI specifics defined); in Sect. 4, the adjusted PMI specific RE model is proposed, which is based on the reused universal RE process parts, but also includes all adjustments required for the successful RE within the scope of PMI; in Sect. 5, the proposed model is applied to the case study example and the model's efficiency is evaluated, based on the results of its application, and future improvements are proposed.

2   PMI Specific RE Model's Quality Criteria

Important factors impacting PMI success are related to PMI knowledge management [32, 35]. There are several effective knowledge management aspects mentioned in related works, including an ability to acquire and apply knowledge and store explicit knowledge for reuse [1, 4, 5, 22]. Additional challenges of PMI are related to the complex and non-structured knowledge, cultural aspects and limited time resources [27, 28]. Knowing that merging companies initially have very limited knowledge about each other, and with additional risks of people leaving the companies during M&A, the ability to acquire tacit knowledge and transform it into usable explicit knowledge plays an important role in PMI success [19]. As PMI covers the whole enterprise architecture and its context, there is an additional need to organize knowledge in such way, that all integration aspects are covered - social, cyber, and physical [7, 8, 11]. Knowledge acquired in PMI is used as a foundation for making decisions on companies' integration, so knowledge should be organized in a way, such that all parties can access and consume it for PMI decision making [15, 16]. With PMI process complexity, it is also hard to extract reusable knowledge for future initiatives to benefit from lessons learned [32, 34]. Ideally, the model should support the accumulation and reuse of knowledge gathered in PMI. Summarizing the above-mentioned issues, the following knowledge management related aspects should be reflected and supported in the RE model:

- Learn about the merging organizations fast and efficiently.
- Create a knowledge structure covering all perspectives: social, cyber, and physical. Both the enterprise architecture and the context should be included.
- Organize the knowledge to provide a solid basis for PMI related decisions, leading to the overall M&A goals; and store all information related to PMI decisions.
- Acquire tacit knowledge and transform it into explicit knowledge.
- Support knowledge reuse across several M&A initiatives.

Created an RE model which, like any model, should use standard modelling notation and be accompanied by a description. Additionally, it should also be possible to effectively apply the RE model during the PMI process, meaning that it should map to the PMI process and domain, should be clear and easy to understand by people involved in the process. As each PMI initiative has its own specifics, the RE model should have an adjustment option. The model also should contain an evaluation part, so that it is possible to assess model application results and improve it in the future. In summary, the RE model should meet the following general model quality criteria [6, 24]:

- Clear language and structure.
- Both static and dynamic perspectives of the PMI initiative are reflected.
- Easy to apply in practice.
- Possible to adjust for specific initiative needs.
- Possible to evaluate the results of the PMI initiative execution.

In the next section we review and evaluate existing models using the defined criteria.

3 Most Suitable Existing RE Models

There are several existing RE models, which are to a larger or smaller extent related to PMI specifics. Further in this section, several of them are inspected and evaluated to see whether the models can fully or partly contribute to PMI specific RE approach.

REAM model [23] describes how a specific domain model and glossary can be created. This model helps to introduce a common domain language. The model uses standard UML class diagram notation. It specifically focuses on a big data domain, however, the methodology itself is universal and could be applied to any other domain, including PMI.

Knowledge structuring model [9] is focused on the PMI knowledge hierarchical structure model and could be used as the PMI domain knowledge classification approach.

Socio-cyber-physical systems (SCPS) domain modelling approach BSCPNet [10] proposes how knowledge in socio-cyber-physical systems can be modelled for decision support purposes. This approach could be applicable also for PMI decisions.

Goal models and a problem frames model [3] could be used for integration of PMI goals into the overall PMI initiative. This model also shows how the context of the specific goal could be investigated and used for the corresponding requirements definition. The model operates with two types of domains – "referred" and "constrained", which are connected through the "machine" satisfying the goal.

KMoS-RE [14] is a comprehensive process framework for aggregation of knowledge management and requirements engineering, based on the SECI framework, but integrated with the RE process. The framework also addresses the problem of acquiring tacit knowledge and transforming it into explicit knowledge. This framework could be used as a foundation for a PMI specific RE model.

Knowledge management framework with a focus on knowledge application – decision making (SOAD) [2] explores how the knowledge gathered can be prepared for the later decision-making process. This approach could be applied in the PMI process for supporting decision making activities related to the integration.

Experience factory [18] defines the model of how the knowledge acquired and produced during the PMI initiative could be stored and accessed later during the same PMI or within the scope of other PMI initiatives. The same question is explored in MIMIR model [5], but the process of knowledge creation is defined on the more detailed level, with specific roles and artefacts defined.

As can be seen in Table 1, none of these frameworks fully addresses all stated requirements for a PMI specific RE model. However, each of them can be incorporated into a PMI specific RE model to be adjusted for the specific needs of PMI.

Table 1. Existing RE model applicability for PMI RE.

RE model	Main theoretical foundation	Supported PMI RE criteria
REAM model	Big data	Supports SCP perspectives
Knowledge structuring model	BABOK, BIZBOK, REBOK, SWEBOK, BPM CBOK	
BSCPNet	Role-based modelling	
Goal models and a problem frames model	KAOS, MSC	Support PMI related decisions
KMoS-RE	SECI	Supports learning merging organizations
Knowledge management framework	SOAD	Supports tacit knowledge acquisition
Experience factory	Experience factory	Support knowledge reuse
MIMIR model	SDLC	

4   PMI Specific RE Model – A Proposal

As was shown in the previous section, there is no one single framework or model that would fully satisfy previously stated goals and quality criteria. In this section, we describe the proposal for a PMI specific RE model, which is based on the PMI process, focused on knowledge management [9, 10, 12], and which is extended with RE activities, adjusted for specific PMI needs. Figure 1 shows the mapping of the phases of the PMI process with the RE activities applied to achieve similar goals as corresponding PMI phases. As can be seen, the initial assessment phase of PMI corresponds to the RE elicitation phase, the PMI planning phase maps to the RE analysis and future state definition phase, and the PMI execution can be perceived as the RE implementation and evaluation phase. From the knowledge management perspective, the PMI initial assessment phase focuses on the acquiring of knowledge about the current state, the planning phase creates knowledge about the future state, and the execution phase accumulates knowledge on the PMI implementation. So, we can look on the PMI process as we do on the RE process, where each phase is focused on a different type of knowledge.

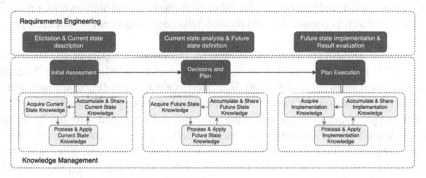

Fig. 1. Relationships between PMI phases and RE activities.

As the model should represent the relationship between RE and knowledge management, each RE phase should be considered from the knowledge perspective.

The following PMI related adjustments are proposed for the initial assessment phase:

- Knowledge should be acquired and structured based on the common PMI domain model.
- Knowledge should be structured hierarchically to highlight PMI goals and requirements.
- All socio-cyber-physical aspects should be covered during the acquisition of knowledge.
- Current architecture and context knowledge should be extracted and documented.
- Tacit knowledge should be acquired together with explicit knowledge.
- Knowledge should be structured and stored in a format that is supporting PMI decisions.

The following PMI related adjustment is proposed for the decisions and planning phase:

- All decisions made should be documented and available for reuse.

The following PMI related adjustments are proposed for the plan execution phase:

- All decisions made should be evaluated.
- Knowledge about new architecture should be documented.

Framework integration requires consolidation of notations. One option is that the following notations can be used for the models:

- For the static perspective of the PMI domain and related artefacts – UML class diagram.
- For the dynamic perspective PMI activities – UML activity diagram.

First, we define a dynamic perspective of the model. Each PMI phase is modeled as a process, which is then divided into three sequential phases – acquire knowledge, process & apply knowledge, and accumulate & share knowledge. For each phase, corresponding RE activities are defined. This paper describes process on quite a high level of abstraction and future research is intended to add more details regarding execution of each process step.

The initial assessment phase is focused on the elicitation part of the RE process (Fig. 2). PMI goals are identified, as well as relevant knowledge sources. After that, explicit and tacit knowledge of PMI goals, as well as a context and company architecture, are acquired. For example, if two finance departments are merged, it should be explained why this merger is required and what is the ultimate goal to be achieved (may be resource optimization, efficiency improvements, etc.) Once the goals are known, finance related architecture parts in both companies (systems used, people involved, processes, etc.) as well as corresponding context parts (partners, customers, competitors, etc.) should be investigated. Based on this knowledge, required PMI activities are defined, scoping the PMI. For each identified PMI activity, in-depth knowledge is accumulated for merging companies, from the perspective of company architecture and context. Then knowledge comparison and evaluation are executed with the purpose of uncovering similarities between merging parties, as well as the strong and weak sides of each of these companies. In the previously mentioned example, the finance department of each company could be evaluated regarding process and resource efficiency. After this, all identified knowledge should be documented to be available during the decisions and planning phase.

During the decisions and planning phase all previously acquired and documented knowledge about PMI goals, context, and architecture is extracted, together with historical PMI knowledge from previous initiatives (Fig. 3). This knowledge is used as the basis for defining required PMI decisions. For each required decision, knowledge is analyzed, then decisions are made, and new architecture is defined. For a finance department merger, it could be decided that a purely manual process in one company should be replaced by an automated process already used in another company. As a last step, all decisions made, as well as new architecture, are documented to be reused later during the execution phase.

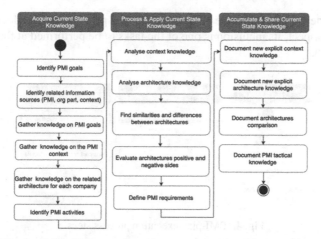

Fig. 2. PMI initial assessment process steps (simplified sequence).

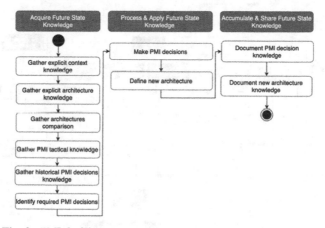

Fig. 3. PMI decisions and plan process steps (simplified sequence).

The last PMI phase is dedicated to the PMI decision execution (Fig. 4). During this phase all stated decisions and new architecture are used as input. Based on those elements, the required change plan is defined and executed. During execution, all results are evaluated, and lessons learned are documented. If there are any changes introduced to the initial plan, these also should be documented and depicted in the updated new architecture.

As can be seen, almost all steps in the described process are linked to some input or output artefacts. To depict this static perspective of the model, we use PMI domain model (Fig. 5). The PMI domain model incorporates PMI knowledge sinks and sources, produced artefacts, as well as all aspects of a company and its context that should be covered in the artefacts. Knowledge related entities are depicted with dark grey color, produced artefacts are depicted with light grey color. White color is used to show knowledge aspects that should be covered or are used as a source for related knowledge or produced artefacts. At this stage of development, the proposed model does not include a multiplicity aspect, its inclusion is one of the tasks to be performed at the later stages of this research.

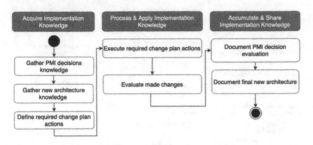

Fig. 4. PMI plan execution process steps.

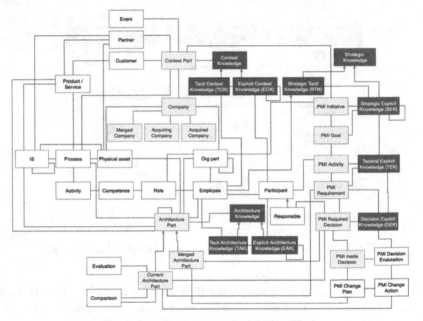

Fig. 5. PMI artefacts data model (simplified).

There are three knowledge aspects that are reflected in the PMI domain model – PMI goal related knowledge, context related knowledge, and company architecture related knowledge. The model covers both current and future company architectures, as well as requirements and change items, justifying the need for the specific PMI initiative. PMI goal related knowledge aims to store evolution of PMI related information, starting with initial goals and then leading to specific decisions and result evaluation. It is important for this part to have a common understanding among the parties involved during a PMI initiative and a reusable template; and also that lessons be learned for the next

PMI initiatives. Knowledge of the current state forms the basis for decision making and supports the analysis of current weaknesses and opportunities in both companies. Context knowledge has the same purpose, as it plays an important role in the PMI decision making process. Knowledge of the future state helps to achieve two parallel goals – (1) define the scope of change and align with the same vision for PMI implementation and (2) create a description of company architecture, which later will become current architecture and can be used for future PMI initiatives. The company's architecture consists of the organization's parts, roles and employees, processes, related physical aspects and information systems. Company context covers customers, partners, and events, that can impact how a company is currently acting. Included entities are already selected in a way that all aspects – social, cyber, and physical, are covered. However, a listed company's architecture and context artefacts can be elaborated in future research and could be transformed into architecture and a context constituent checklist for any PMI initiative. PMI related artefacts are structured hierarchically, starting with the initial PMI vision, leading sequentially to the more detailed PMI goals, activities, requirements and related decisions. Each requirement and decision is linked to the current architectures of the merging companies and future architecture of the merged company. Three process models (covering PMI dynamic perspective) together with one data model (covering PMI static perspective) form the PMI specific requirements engineering model, which can be used as a checklist during knowledge elicitation and knowledge structuring.

5 PMI Specific RE Model – Demonstration of Applicability of the Proposed Model

We validate the proposed model on one example of a PMI initiative. The case discussed in this section is only part of a larger PMI project, but it is still applicable to the overall model illustration and initial validation. The example is based on a real PMI case; however, the case here is simplified to highlight only the most important elements and it is anonymized for confidentiality reasons. In this specific case, two companies A and B are merged. The case covers only part of this merger – the finance merger.

PMI dynamic perspective is represented by three processes according to the PMI process models proposed in the previous section. In order to better highlight relations between dynamic and static perspectives, in Fig. 6, Fig. 7, Fig. 8 PMI processes are depicted accompanied by corresponding PMI artefacts.

During the initial assessment phase, PMI goals were identified (Fig. 6). One of them was the alignment of processes and resource optimization. Based on this initial goal, explicit and tacit goal related knowledge was investigated, based on the PMI vision and interviews with related finance employees, as well as the context and architecture knowledge – finance partners, customers, finance processes, related organization struc-tures, information systems, etc. Based on this investigation, more precise PMI activity was defined – consolidate invoicing process. After this, the invoicing context and related architecture parts were analyzed in both companies. Invoicing architecture consists of the organizational units, processes, required competences and employees involved, as well as information systems used and physical assets. However, as stated previously, the information system aspect should be treated with a higher priority and should form the basis for all other integration aspects. The invoicing context is mainly customers, who partially overlap between companies. For both companies there is also the same regulator organization. For the company A there is also a vendor, providing invoicing system. As a result of the analysis, architecture comparison and evaluation were executed. The invoicing in the company A is a fully automated process; efficient, but not adjusted for specific customer needs. In the company B, invoicing is a manual process; however it is configured for different customers. An important aspect is overlapping customers. One of the requirements stated that, from the customer perspective, in the future there should be one common invoicing approach for both acquired companies. It is important to note, that in the real-life scenario, several requirements were defined, but in the research study we selected only one of them for illustrative needs. As the final step of the initial assessment, all knowledge gathered was documented: invoicing context in both companies, current processes, process comparison and PMI goals for reporting consolidation.

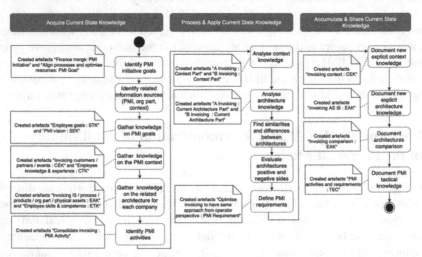

Fig. 6. Reporting PMI initial assessment process steps and produced artefacts (PMI process linked with PMI artefacts).

Knowledge gathered was used as an input for the decisions and planning phase (Fig. 7). Important additional input was the historical knowledge from previous PMI initiatives. In our case this knowledge was lessons learned after previous finance transformations or other department mergers. Based on the PMI goal to create consistent invoicing experience for customers and on all of the knowledge about invoicing architecture and context, the required PMI decisions were defined. In our case only one decision is selected for future analysis – define future invoicing process. Justified by current invoicing comparison and evaluation for both companies, one of the decisions made was to take the invoicing process in the company A as a basis and integrate into it required changes to support invoicing context from the company B. Together with this decision, future invoicing architecture was defined. In the end, all decisions made and corresponding future architectures were documented to be used during the execution phase.

During the execution phase the knowledge about decisions made and future architecture was used as an input (Fig. 8). This knowledge was used to define the required change plan and change actions. For invoicing consolidation, it was decided to merge two finance departments into one common department, as well as to use an automated invoicing process in the future. One of the required change actions was the necessity to train employees from the company B to be able to execute the automated invoicing process. Here again, in the real-life example, the change plan was a detailed artefact, not just the several actions mentioned here. As the next step, the defined change plan was executed, and future architecture was implemented. Together with this, executed changes were evaluated, and corrected. All final decisions, together with the evaluation, as well as final invoicing architecture, were documented as explicit knowledge. This explicit knowledge can be used in future PMI initiatives, related to the finance merger (for example, if another company will be acquired).

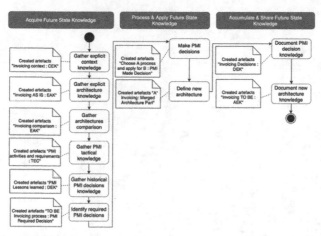

Fig. 7. Reporting PMI decisions and plan process steps and produced artefacts (PMI process linked with PMI artefacts).

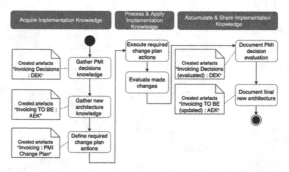

Fig. 8. Reporting PMI plan execution process steps and produced artefacts (PMI process linked with PMI artefacts).

The artefact data model amalgamates the created artefacts and acquired documented explicit knowledge (Fig. 9). The model has the following parts:

- PMI initiative, goals, activities and requirements.
- Company A's and Company B's current invoicing architecture & context.
- Invoicing comparison and evaluation.
- PMI decisions made, related change plan and future invoicing architecture.
- Architecture, context, and PMI initiative related tacit and explicit knowledge.

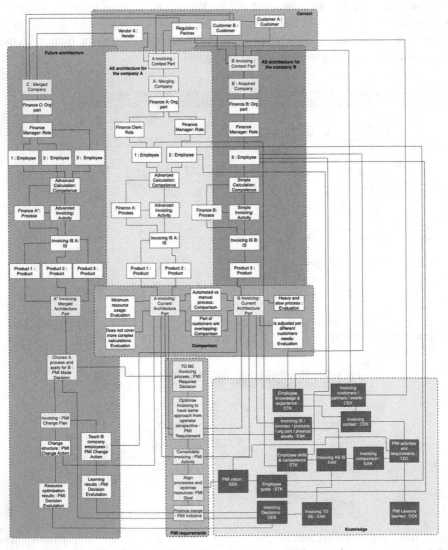

Fig. 9. Reporting PMI artefacts data model example.

Invoicing architectures incorporate the information about the products for which invoicing is produced, information systems used, the process itself, as well as required executer competences and employees involved together with their roles. Invoicing contexts incorporate the information about customers using the resultant invoicing, the regulator constraining the process, partners contributing to this process, as well as the vendor supporting invoicing information system.

6 Summary

This research focuses on the RE within the scope of PMI initiatives. With knowledge management being an important part of the PMI, the assumption that PMI RE should be adjusted to support PMI knowledge management needs is explored. Existing RE models are explored and evaluated from the knowledge management support perspective. As none of them has qualified as the holistic approach, a new PMI specific RE model is proposed and demonstrated through practical application. At this stage of the research, the main focus is on a generic RE process model, and an additional effort is required to transform it into a detailed RE model. Nevertheless, the proposed model already meets the majority of stated quality criteria. The model shows how to integrate tacit knowledge and the PMI domain into standard RE approaches, which has not been highlighted in any previous research. Compared to other existing RE approaches explored in this paper, the proposed PMI specific RE model covers both dynamic and static perspectives and has been demonstrated to be applicable in real life PMI initiatives. It also uses a standard UML notation, widely used in modelling.

In further steps of this research, we intend to test the model in other real cases and elaborate it further. For instance, a more detailed description of each process step, and smaller granularity descriptions of created knowledge artefacts are intended to be presented. The model should be expanded to include adjustment mechanisms for its application in various PMI initiatives. Enhanced knowledge validation and verification activities are intended to reduce the risk that acquired knowledge is outdated.

References

1. Wilson, L.T., Snyder, C.A.: Knowledge management and IT: how are they related? IT Professional **1**(2), 73–75 (1999). https://doi.org/10.1109/6294.774944
2. Prasetyo, N.A., Bandung, Y.: A design of software requirement engineering framework based on knowledge management and Service-Oriented Architecture Decision (SOAD) modeling framework. In: 2015 International Conference on Information Technology Systems and Innovation, ICITSI 2015 - Proceedings (2016). https://doi.org/10.1109/ICITSI.2015.7437708
3. Mohammadi, N.G., Alebrahim, A., Weyer, T., Heisel, M., Pohl, K.: A framework for combining problem frames and goal models to support context analysis during requirements engineering. In: Cuzzocrea, A., Kittl, C., Simos, D.E., Weippl, E., Xu, L. (eds.) CD-ARES 2013. LNCS, vol. 8127, pp. 272–288. Springer, Heidelberg (2013). https://doi.org/10.1007/978-3-642-40511-2_19
4. Baxter, D., et al.: A framework to integrate design knowledge reuse and requirements management in engineering design. Robot. Comput.-Integr. Manuf. **24**(4), 585–593 (2008). https://doi.org/10.1016/j.rcim.2007.07.010

5. Rocha, Á., Correia, A.M., Adeli, H., Reis, L.P., Teixeira, M.M.: Preface. In: Advances in Intelligent Systems and Computing, vol. 445, pp. V–VI (2016). https://doi.org/10.1007/978-3-319-31232-3

6. Taylor, C., Sedera, W.: Defining the quality business process reference models. In: 14th Australasian Conference on Information Systems, vol. 196, pp. 1–10 (2003). http://aisel.ais net.org/acis2003/79

7. Rehman, S., Gruhn, V., Shafiq, S., Inayat, I.: A systematic mapping study on security requirements engineering frameworks for cyber-physical systems. In: Wang, G., Chen, J., Yang, L.T. (eds.) SpaCCS 2018. LNCS, vol. 11342, pp. 428–442. Springer, Cham (2018). https://doi.org/10.1007/978-3-030-05345-1_37

8. Haidar, H., Kolp, M., Wautelet, Y.: An integrated requirements engineering framework for agile software product lines. In: van Sinderen, M., Maciaszek, L.A. (eds.) ICSOFT 2018. CCIS, vol. 1077, pp. 124–149. Springer, Cham (2019). https://doi.org/10.1007/978-3-030-29157-0_6

9. Aoyama, M.: Bridging the requirements engineering and business analysis toward a unified knowledge framework. In: Link, S., Trujillo, J.C. (eds.) ER 2016. LNCS, vol. 9975, pp. 149–160. Springer, Cham (2016). https://doi.org/10.1007/978-3-319-47717-6_13

10. Smirnov, A., Sandkuhl, K.: Context-Oriented knowledge management for decision support in business socio-cyber-physical networks: conceptual and methodical foundations. In: Conference of Open Innovation Association, FRUCT, April 2017, pp. 413–419 (2017). https://doi.org/10.23919/FRUCT.2017.8071342

11. Kirikova, M.: Continuous requirements engineering in the context of socio-cyber-physical systems. In: Robal, T., Haav, H.-M., Penjam, J., Matulevičius, R. (eds.) DB&IS 2020. CCIS, vol. 1243, pp. 3–12. Springer, Cham (2020). https://doi.org/10.1007/978-3-030-57672-1_1

12. Chang, C.H., Lu, C.W., Chu, W.C.: Improving software integration from requirement process with a model-based object-oriented approach. In: Proceedings - The 2nd IEEE International Conference on Secure System Integration and Reliability Improvement, SSIRI 2008, pp. 175–176 (2008). https://doi.org/10.1109/SSIRI.2008.39

13. Nguyen, T.H., Vo, B.Q., Lumpe, M., Grundy, J.: KBRE: a framework for knowledge-based requirements engineering. Software Qual. J. **22**(1), 87–119 (2013). https://doi.org/10.1007/s11219-013-9202-6

14. Olmos, K., Rodas, J.: KMoS-RE: knowledge management on a strategy to requirements engineering. Requirements Eng. **19**(4), 421–440 (2013). https://doi.org/10.1007/s00766-013-0178-3

15. Hirani, H., Ratchev, S.: Knowledge-based requirements engineering for reconfigurable precision assembly systems. In: Camarinha-Matos, L.M. (ed.) BASYS 2004. IIFIP, vol. 159, pp. 359–366. Springer, Boston (2005). https://doi.org/10.1007/0-387-22829-2_38

16. Pavanasam, V., Subramaniam, C., Mulchandani, M., Parthasarathy, A.: Knowledge based requirement engineering framework for emergency management system. In: 2010 2nd International Conference on Computer Engineering and Applications, ICCEA 2010, vol. 1, pp. 601–605 (2010). https://doi.org/10.1109/ICCEA.2010.123

17. Mishra, D., Aydin, S., Mishra, A., Ostrovska, S.: Knowledge management in requirement elicitation: Situational methods view. Comput. Stand. Interfaces **56**, 49–61 (2018). https://doi.org/10.1016/j.csi.2017.09.004

18. Aurum, A., Parkin, P., Cox, K.: Knowledge management in software engineering education. In: Proceedings - IEEE International Conference on Advanced Learning Technologies, ICALT 2004, June, pp. 370–374 (2004). https://doi.org/10.1109/ICALT.2004.1357439

19. Serna, E., Bachiller, O., Serna, A.: Knowledge meaning and management in requirements engineering. Int. J. Inf. Manage. **37**(3), 155–161 (2017). https://doi.org/10.1016/j.ijinfomgt.2017.01.005

20. Bresciani, P., Donzelli, P., Forte, A.: Requirements engineering for knowledge management in eGovernment. In: Wimmer, M.A. (ed.) KMGov 2003. LNCS, vol. 2645, pp. 48–59. Springer, Heidelberg (2003). https://doi.org/10.1007/3-540-44836-5_5
21. Chu, M., Tian, S.: Research on knowledge management of collaborative design. In: Proceedings of the International Conference on E-Business and E-Government, ICEE 2010, pp. 1890–1893 (2010). https://doi.org/10.1109/ICEE.2010.478
22. Yang, X.: To facilitate knowledge management using basic principles of knowledge engineering. In: KESE 2009 - 2009 Pacific-Asia Conference on Knowledge Engineering and Software Engineering, pp. 94–97 (2009). https://doi.org/10.1109/KESE.2009.33
23. Arruda, D., Madhavji, N. H.: Towards a big data requirements engineering artefact model in the context of big data software development projects: poster extended abstract. In: Proceedings. 2017 IEEE International Conference on Big Data, Big Data 2017, January(v) 2018, pp. 4725–4726 (2017). https://doi.org/10.1109/BigData.2017.8258521
24. Krogstie, J., Lindland, O., Sindre, G.: Towards a deeper understanding of quality in requirements engineering. In: Iivari, J., Lyytinen, K., Rossi, M. (eds.) CAiSE. LNCS, vol. 932, pp. 82–95. Springer, Heidelberg (1995). https://doi.org/10.1007/3-540-59498-1_239
25. Katina, P.F., Keating, C.B., Jaradat, R.M.: System requirements engineering in complex situations. Requirements Eng. **19**(1), 45–62 (2012). https://doi.org/10.1007/s00766-012-0157-0
26. Szejka, A.L., Aubry, A., Panetto, H., Júnior, O.C., Loures, E.R.: Towards a conceptual framework for requirements interoperability in complex systems engineering. In: Meersman, R., et al. (eds.) OTM 2014. LNCS, vol. 8842, pp. 229–240. Springer, Heidelberg (2014). https://doi.org/10.1007/978-3-662-45550-0_24
27. Safavi, M.: Advancing post-merger integration studies: a study of a persistent organizational routine and embeddedness in broader societal context. Long Range Plann., 102071 (2021). https://doi.org/10.1016/j.lrp.2021.102071
28. Riedel, L., Asghari, R.: Mergers and acquisitions as enabler of digital business transformation: introducing an integrated process. In: Sangwan, K.S., Herrmann, C. (eds.) Enhancing Future Skills and Entrepreneurship. SPLCEM, pp. 271–279. Springer, Cham (2020). https://doi.org/10.1007/978-3-030-44248-4_27
29. Escalona, M.J., Aragón, G., Linger, H., Lang, M., Barry, C., Schneider, C. (eds.): Information System Development. Springer, Cham (2014). https://doi.org/10.1007/978-3-319-07215-9
30. Wirz, P., Lusti, M.: Information technology strategies in mergers and acquisitions: an empirical survey. In: Proceedings of the Winter International Synposium on Information and Communication Technologies (WISICT), vol. 58, pp. 1–6 (2004). http://portal.acm.org/citation.cfm?id=984720.984777
31. Hedman, J., Sarker, S.: Information system integration in mergers and acquisitions: research ahead. Eur. J. Inf. Syst. **24**(2), 117–120 (2015). https://doi.org/10.1057/ejis.2015.2
32. Kumar, J., Haine, P., Brown, M.: Design leadership for mergers and acquisitions. Interactions **21**(3), 74–76 (2014). https://doi.org/10.1145/2594208
33. Tikhomirov, A.F., Kalchenko, O.A., Bogacheva, T.V.: Assessment of the synergistic effect from solvency changes in M&A transactions based on a dynamic model. In: ACM International Conference Proceeding Series (2019). https://doi.org/10.1145/3372177.3373282
34. Motta, M., Peitz, M.: Big tech mergers. Inf. Econ. Policy **54**, 100868 (2021). https://doi.org/10.1016/j.infoecopol.2020.100868
35. Katz, M.L.: Big Tech mergers: Innovation, competition for the market, and the acquisition of emerging competitors. Inf. Econ. Policy **54**, 100883 (2021). https://doi.org/10.1016/j.infoecopol.2020.100883

Enterprise Modeling Methods
and Frameworks

*its*VALUE - A Method Supporting Value-Oriented ITSM

Henning Richter[ID] and Birger Lantow[✉][ID]

University of Rostock, 18051 Rostock, Germany
{henning.richter,birger.lantow}@uni-rostock.de

Abstract. In 2020, the new ITIL 4 standard was introduced. ITIL standardization had and still has a big influence on how IT-Service Management is seen and performed in practice. Thus, the new standard is expected to have a high impact as well. A key element of ITIL 4 is the strong focus on Stakeholder Value in the analysis of IT-Services. Yet apart from ITIL, stakeholder orientation is a current trend in business analysis. *its*VALUE method and Modeller provide means to model and analyze value delivery in IT-Services and thus can be used Service Design in an ITIL 4 context. It combines "traditional" approaches to value stream analysis and service modelling and adds concepts and functionalities that meet the requirements of IT-Service Management and ITIL 4. This work introduces the basic concepts of *its*VALUE and discusses a first case study based evaluation and its results. Generally, the participants involved perceived the method as suitable to support value-oriented IT-Service Management.

Keywords: ITIL · Service modelling · Service value · Value stream modelling · Stakeholder value · Service blueprinting · IT-Service management

1 Introduction

A new era in the field of IT has begun, as services are the biggest and most dynamic market component of both industrial and developing countries [3]. Moreover, services are the most important goods for generating organizational value for both the company itself and its customers. Further, almost any current service is supported by IT components, and IT is developing as fast as never before in human history. Thus, companies can take advantage of enhancing their understanding and performance for IT-Service Management. New techniques (e.g. cloud computing, machine learning, blockchain, etc.) enabled new opportunities for the value chains and value creation of companies. Thus, IT (especially IT-Service Management) is one of the most important business drivers companies should carefully consider nowadays to achieve competitive advantage. ITIL v3 is a well-known reference for best practices in IT-Service Management. It describes processes, roles, and KPIs. Current trends like increasing market dynamics, the

© Springer Nature Switzerland AG 2021
R. A. Buchmann et al. (Eds.): BIR 2021, LNBIP 430, pp. 133–149, 2021.
https://doi.org/10.1007/978-3-030-87205-2_9

advent of agile software development, and the integration of products and services made a revision necessary. In 2020 ITIL 4 was released. It primarily focuses on enabling responding to new stakeholder demand quickly and simply. According to [3], a company's purpose is to create value for its stakeholders. Everything a company does must serve (directly or implicitly) the creation of value for its stakeholders. ITIL has a strong industrial background and it is likely that many companies will adopt the new version in order to improve their IT-Service Management capabilities. While ITIL 4 generally describes these capabilities and their integration, a concrete method or tool-set for the integration of stakeholder value in service design is not provided. Even if an enterprise does not intend to implement ITIL 4, considering Stakeholder Value in IT-Service Management can improve demand orientation.

A literature analysis [14] showed that there are approaches like IT S-SB [16] or VSD 4.0 [8,9] that support modelling and analysis of IT-Service delivery from a value and stakeholder-oriented perspective. However, these are either specialized on a certain use case (e.g. IT S-SB) or just miss some aspects and requirements of value delivery modelling that are important to ITIL v4. Furthermore, there are notations like Value Delivery Modelling Language[1] (VDML) and the Archimate Motivation Extension[2] that provide necessary modelling concepts. Yet, there is no method support in terms of procedures and guidelines for the creation and usage of models. Usage of these notations for value-oriented IT-Service Management would also imply a further operationalization since they remain on a high abstraction level.

"It's a Value Added Language You Employ" (*its*VALUE) has been developed to provide a method and a tool (the *its*VALUE Modeller) for modelling and analysing IT-Service value delivery that can be generally applied to IT-Services and build on proven concepts of Service-, Value, and Enterprise Modelling. This study describes the method components and the general process of *its*VALUE application in Sect. 2. The meta-model describing the model notation and forming a base for modelling tool implementation is presented in Sect. 3. We discuss a first case study based evaluation of the approach in Sect. 4. The last Sect. 5 describes the current state of the development and provides an outlook to next steps to foster *its*VALUE application.

2   The *its*VALUE Method

As mentioned in the introduction, *its*VALUE combines and amends proven concepts of Service-, Value, and Enterprise Modelling in order to support Value Stream analysis for IT-Services especially with a focus on ITIL 4 [3] because of its practical relevance. According to the ITIL 4 documentation, [1,3], *value* is a set of a perceived usefulness, importance and benefits of something. This goes beyond "traditional" value stream analysis and value modelling, where value stream optimization is focused on processing times and modelling the exchange

---

[1] https://www.omg.org/spec/VDML/.

[2] https://pubs.opengroup.org/architecture/archimate3-doc/.

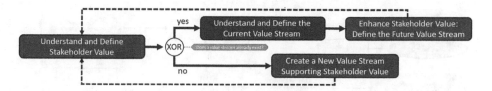

Fig. 1. *its*VALUE framework

of economic value. Further, ITIL 4 recommends Service Blueprinting to model and understand the customer journey. This helps to analyse the stakeholder perspective in a service setting. Yet, the "traditional" Service Blueprinting approach does not consider the complexity of IT-Services. We developed *its*VALUE to provide a sound combination of new ideas and requirements for value-oriented IT-Service Management based on ITIL 4 and the named "traditional" approaches. Thus, key elements of *its*VALUE are taken from VSA 4.0, VSM 4.0, VSD 4.0 [8,9,11,12] and VSMN [10] (as extensions of "traditional" value stream modelling considering information processing and stakeholder perspectives), IT S-SB [16] (as an approach adding information technology to "traditional" Service Blueprints) and 4EM [15] (as a participatory enterprise modelling method supporting the integration of stakeholders and provides concepts that allow to model context influence on value delivery).

The *its*VALUE method consists of four components that describe the steps to model, analyse, and (re-)design value streams of IT-Services. Figure 1 shows the method framework, aligning the components in a process. Two alternatives are distinguished depending on whether there is already an existing value stream or not. As shown in the figure, we understand that the method has to follow an iterative approach in order to develop the required insight into the analysed value stream. Due to Geist [6] gathering promising results when performing four iterations with stakeholders in a modelling context, we recommend four iterations for *its*VALUE as well, especially when starting with an entirely new service. The following Sects. 2.1, 2.2, 2.3 and 2.4 introduce each component of *its*VALUE in detail.

2.1 Understand Stakeholder Value (SV)

The first phase of our *its*VALUE framework focuses on understanding and explicitly defining what each relevant stakeholder affected by a service actually values. Besides the customers, [2] lists employees, managers, suppliers, partners, the media, the public, and much more as important stakeholders a company should also consider. It is recommended to explore and understand the needs of each relevant stakeholder to define what they value. The results of these investigations are modeled in the Value Perception Model (VPM). A VPM is created for each relevant stakeholder. In these models, the stakeholder's value assumptions are placed around the stakeholder.

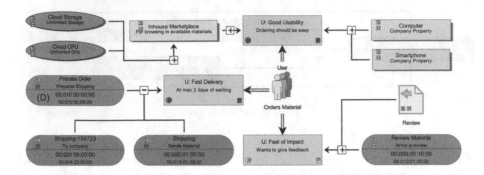

Fig. 2. Exemplary VPM of a User

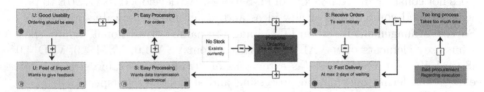

Fig. 3. Exemplary SVM for different stakeholders including the User from Fig. 2

Each modelled StakeholderValue should be classified regarding its Relevance Factor (RF) and Current Performance Level (CPL). For the RF, we defined four rating-values: *Mandatory*, *Moderate*, *High* and *Outstanding*. Here, we adopted the idea of distinguishing between hygiene and success factors from [13]. A stakeholder value is considered as a hygiene factor if it causes dissatisfaction when missing while not providing much potential to increase satisfaction when delivered. Oppositely, we understand *Moderate*, *High* and *Outstanding* Stakeholder-Values as different levels of relevance for success factors: If success factors are provided, they increase satisfaction (depending on relevance) while causing not much dissatisfaction when missing.

Current Performance Levels reflect the perceived performance of value delivery. Generally, the CPL can only be assessed, if a Value Stream or its components are already implemented. However, our method requires a CPL for each Stakeholder Value to perform a value stream analysis. Thus, a default CPL should be defined for new Value Streams and new Value Stream components. For the CPL, we define the following rating-values: *Poor*, *Moderate*, *High* and *Outstanding*. We recommend to use *Moderate* as the default value for the CPL, as this setting does not influence value stream analysis outcomes. An exemplary VPM can be seen in Fig. 2.

Once each relevant stakeholder's value perception is explicitly defined, the dependencies and relations among all Stakeholder Values should be investigated. We understand Stakeholder Value as a specialization of business goals in the sense of 4EM by Sandkuhl et al. [15]. Thus, delivering different values may cause mutual support, obstruction, or even contradiction. Results are reflected in the

Stakeholder Value Map (SVM). Understanding how different Stakeholder Values affect each other can assist companies in detecting which SV are beneficial or problematical to other Stakeholder Values. Combined with the RF and CPL of a value, such an understanding can be useful for deciding which Stakeholder Value is more important than another when defining development actions. Figure 3 shows an exemplary SVM.

2.2   Understand and Model the Current Value Stream (CVS)

If a Value Stream already exists, it must be modeled, understood, and analysed before it can be improved. The second component of our *its*VALUE framework deals with these issues. Mainly, it is divided into two clusters of activities each serving a different purpose: the value stream modelling followed by the value stream analysis.

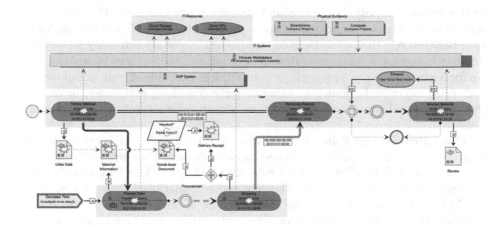

Fig. 4. Exemplary VSB

Value Stream Modelling. Firstly, any process, activity that is part of the Current Value Stream should be collected and ordered from the initial demand to the final delivery (1st step of Value Stream Analysis 4.0). We developed the Value Stream Blueprint (VSB) for the visual representation of the value stream. Basically, it is a combination of IT Self-Service Blueprint and Value Stream Model and Notation that can be used to model Value Streams. Figure 4 shows an exemplary VSB.

Secondly, all additional components (e.g. storage media, IT systems, physical evidence, information, waiting times, etc.) should be collected and connected to the processes and activities they are used or required at (2nd and 3rd step of Value Stream Analysis 4.0). This step can also be carried out by using a VSB.

Thirdly, required data for each process or activity should be defined (4th step of Value Stream Analysis 4.0). Therefore, the following information of each desired or captured data should be defined: *desired (yes/no)*, *captured (yes/no)*, *used (yes/no)* and *acquisition (automatically/manually)*. Collecting these information for each data of each process or action allows calculating Key Performance Indicators for value stream analysis.

Fourthly, we advice mapping all components of the Current Value Stream to the VPMs created earlier. The VPMs describe Stakeholder Values belonging to their stakeholder. To map these VPMs with all components that are part of the Current Value Stream, each component should be analysed with regard to its actual effects on any Stakeholder Value. This identification provides awareness and understanding of the Current Value Stream and Stakeholder Values. It is possible to detect hidden, indirect effects of components in the value stream on the value perceptions as well as new value perceptions. While new value perceptions should be added to the respective VPM, indirect effects will be further investigated in the analysis step.

Value Stream Analysis. In the analysis step, waste inside the Current Value Stream and potential improvement spots are detected (step 5 and 6 from VSA 4.0). Besides cycle time analysis from "traditional" value stream analysis, we distinguish between the analysis of data processing and the analysis of stakeholder perspectives. Generally, the analysis concentrates on value stream components that cause poor performance in value delivery (e.g. long idle times, increased effort, or general obstacles) and are thus considered wasteful or as producing "Waste".

Data Processing: Following Meudt et al. [11,12], data can generate waste regarding its usage, acquisition, processing, and storage. They originally derived the Digitization Rate (DR), Data Availability (DA) and Data Usage (DU) Key Performance Indicators (KPIs). The required information for calculation (see Fig. 5) is part of the Current Value Stream model. If the DR is lower than 1, the process has the potential to become more automatized and thus more efficient. If the DA is not equal to 1, the process receives either not enough required or too much unnecessary data. If the DU is lower than 1, the process captures unnecessary data. To conclude, with every KPI close or even equal to 1, a process produces very low or even no waste [9,12].

Stakeholder Perspectives: Following ITIL 4, each activity of a Value Stream should generate more value than it consumes. This originates from an economic perspective of value. Here, consumed value means costs that can be calculated for value delivery activities. Waste would be negative revenue from an activity. Especially in the service domain, delivered value can also be intangible and non-economic and thus generally not quantifiable in terms of costs and revenues. The VPMs mostly describe this "unquantifiable" type of value. Hence, *its*VALUE does not focus on the detection of negative revenues. Instead, we concentrate on the identification of the analysis and detection of waste based on RF, CPL, and dependencies in the VPMs. We developed an algorithm detecting for each component of a Value Stream how it supports and hinders Stakeholder Values. With the Supporting Score (SS) and Hindering Score (HS) of such a

$$DR = \frac{\sum \textit{automatically acquired (and digitally captured) data}}{\sum \textit{captured data}}$$

$$DA = \frac{\sum \textit{captured data}}{\sum \textit{desired data}} \qquad\qquad DU = \frac{\sum \textit{used data}}{\sum \textit{captured data}}$$

Fig. 5. Calculating the DR, DA & DU by [9,12]

$$RF : \begin{cases} 3 & \textbf{if} \quad "Mandatory" \\ 1 & \textbf{if} \quad "Moderate" \\ 2 & \textbf{if} \quad "High" \\ 3 & \textbf{if} \quad "Outstanding" \end{cases} \quad CPL : \begin{cases} -3 & \textbf{if} \quad "Poor" \\ 1 & \textbf{if} \quad "Moderate" \\ 2 & \textbf{if} \quad "High" \\ 3 & \textbf{if} \quad "Outstanding" \end{cases} \quad IF : \begin{cases} \frac{1}{2} & \textbf{if} \quad "Low" \\ 1 & \textbf{if} \quad "Moderate" \\ \frac{3}{2} & \textbf{if} \quad "High" \end{cases}$$

$$VW = RF \times CPL$$

Fig. 6. Variable assignments and calculating the value weight of a stakeholder value

component, we defined two new KPIs addressing this issue. The SS provides the created Stakeholder Value attributed to a Value Stream component, while the HS provides the waste attributed to that component. Thus, it is possible to identify components creating much value as well as components creating no value at all and as well components that create waste by nonperformance or hindering the creation of value. The calculation is based on the Value Weight (VW) as shown in Fig. 6. Not considering economic values, we are not dealing with a metric scale. Thus, direct weighting of the created value against the detected waste of negative influences on the delivery of certain Stakeholder Values problematically. Consequently, the interpretation of both is left to human analysis. In addition to the ratings of the VPM, an Impact Factor (IF) of a component affecting service delivery can be specified for fine-tuning. Based on the VW, the SS and HS are calculated considering direct and indirect supports and hinders relations in the VPM.

Ranking of Potential Improvement Spots: The previously described KPIs support identifying potential improvement spots. Further, key Stakeholder Value can also be identified showing which Stakeholder Value or key components require enhancements the most. E.g., those Stakeholder Values having a low CPL and high RF embody a high potential for improvement. Therefore, the SS and HS of all affecting components can be considered as well. If they do not support those Stakeholder Values or even hinder them, they embody a high potential of improvement. This may reveal, which specific development actions are required to directly enhance any Stakeholder Value. Moreover, we suggest to consider the SVM earlier created as well, as it might assist in not missing out any problematic dependencies or relations between several Stakeholder Values. Understanding those relations can support identifying the best development actions to enhance as much Stakeholder Values as best as possible. Further, the data processing KPIs can be used to identify media breaches and other problems in the data supply.

2.3 Enhance Stakeholder Value: Define the Future Value Stream (FVS)

Once the Current Value Stream is understood and explicitly defined, one or multiple potential Future Value Streams can be developed. For this purpose, we derived our approach in reference to [11,12,15]. The SVM can be used to define further information like opportunities, problems, constraints, or causes to sharp the general scope of improvement to be planned. Thus, opportunities regarding bad performing Stakeholder Values should be defined in the SVM. Development actions explicitly affecting components that affect Stakeholder Value could be defined in the VPMs.

Additionally, VSD 4.0 [11,12] recommends maximizing the flow of the Value Stream, especially by avoiding as much waiting times as possible. For information flows, avoiding media breaches seems to be the most important task. As media breaches require a manual processing of the information from one media to another, they immediately prevent a continuous and uninterrupted flow of information. The general solution to avoid media breaches is the development of machine-to-machine interfaces. The re-design of complex processes in the Value Stream should be done in a top-down approach (i.e., always considering the corresponding outcomes that result for the entire Value Stream), as combinations of optimised components does not inevitably lead to an optimized Value Stream. Following ITIL 4, an optimized Value Stream is much more important than a set of highly optimized processes. Still, *Decomposition* of processes is possible in *its*VALUE to support a more specific local view. For each process, development actions should be defined to satisfy each process needs regarding Digitization Rate, Data Availability & Data Usage. This also includes a thorough analysis of information demands and determining how required data and information is stored and accessed.

Lastly, the Future Value Stream is modeled. For *its*VALUE, this implies not just modelling the VSB but also future versions of all VPMs and the SVM. After modelling the Future Value Stream, a new iteration of the *its*VALUE method is recommended to find unwanted side effects and iterative refinement of the Value Stream. However, we advise revising each service modeled with *its*VALUE on a regular basis, as [3] underlines the importance of achieving high business flexibility. Thereby, a company should become capable of adapting to rapidly changing demands and requirements and to sustainable satisfy all stakeholder's needs and desires.

2.4   Create an Initial Value Stream Supporting Stakeholder Value

In Service Design, it might be possible that a new service has to be developed. This includes modelling a Future Value Stream and connected VPMs and SVMs. The first step of this method component is a combination of the 1st steps of the components "Understand Stakeholder Value" and "Enhance Stakeholder Value" - the definition of value opportunities and the collection of all processes required for the Value Stream. Each potential process of the new Value Stream should be immediately evaluated regarding its effects on the defined Stakeholder Values and its data processing KPIs. If a process is considered as not beneficial to any Stakeholder Value or as hindering the continuous flow of the Value Stream, a company should drop that process or activity directly. This analysis is performed in the next steps.

The second step of this method component focuses on a similar domain to the 2nd step of the "Enhance Stakeholder Value" component. Potential waiting time should be eliminated or at least minimized, as they embody waste by decreasing the efficiency and even potentially effectiveness of a Value Stream in the sense of [11]. The leaner and smarter a Value Stream is, the more efficient it performs.

The third step of this method component combines the 3rd step of "Understand Stakeholder Value" (define data requirements) and the 4th step of "Enhance Stakeholder Value" (development actions for data requirements). Once each required process or activity has been identified, the data and information requirements for each one of them should be defined to immediately achieve good KPI values (see Sect. 2.2).

For further refinement, the fourth step of this method component is a combination of the 2nd step of "Understand Stakeholder Value" (identifying supporting components) and the 3rd step of "Enhance Stakeholder Value" (development actions to avoid media breaches). Reasonable and effective (i.e. providing a supportive influence on Stakeholder Values) supporting components should be identified and connected to all processes or activities requiring or dealing with data.

Lastly, a combination of the 5th steps of "Understand Stakeholder Value" (mapping of the Value Stream components and Stakeholder Values) and "Enhance Stakeholder Value" (modelling the future state) should be performed to explicitly define how each Stakeholder Value should be affected and perform with this new Value Stream. Like already argued at the end of Sect. 2.3, this should be checked by entering and performing a new iteration of *its*VALUE. Especially, a reconsideration of what highly relevant stakeholders value should be performed to ensure not missing out any forgotten or even new value perceptions that are important for the service and its value creation (like already argued at the end of Sect. 2.2).

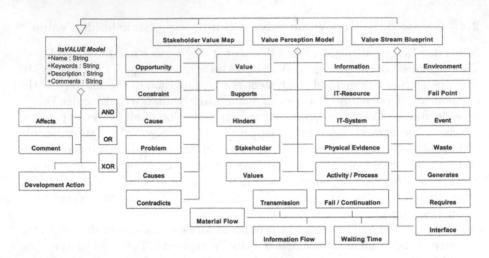

Fig. 7. *its*VALUE meta model: model types

3 The *its*VALUE Notation and Modeller

Figure 7 shows the modelling concepts that can be used in *its*VALUE and the sub-models (VPM, SVM, and VSB) where they can be used. For reasons of brevity we omitted the complete meta model of *its*VALUE. In addition to Fig. 7 there is also a definition of the syntax and the attributes of the *its*VALUE notation that form the meta model. For the implementation of the modelling tool, a visual notation in addition to this abstract notation has been developed. An impression of the attributes, possible relations and the visual elements is given in the method description and the provided example models (cf. Sect. 2). The *its*VALUE modeller[3] has been developed on the ADOxx platform[4]. The prototypical implementation provides all notation elements and sub-models of the meta-model and functionalities for value stream analysis as described in Sect. 2.2.

4 Case Study

A case study has been performed in order to evaluate *its*VALUE. This evaluation has been done based on an *its*VALUE modelling session followed by guided interviews with the participants. We divided the interviews into four parts in order to be able to differentiate method artefacts with regard to their perception and their influence on the method perception. In this paper, we focus on the results for the method framework and the concepts. Additional artefacts that have been evaluated are the visual notation and the modelling tool. Since the

---

[3] https://www.omilab.org/activities/projects/details/?id=148.
[4] https://www.adoxx.org.

Table 1. Mapping the participants (P) to the groups of stakeholders involved

Group of Stakeholders	P A	P B	P C	P D	$\sum$
User (Employee)	*Yes*	*Yes*	*Yes*	*Yes*	4
Global Service Desk	*No*	*No*	*No*	*No*	0
Team Leader	*Yes*	*No*	*No*	*No*	1
Client Service Team	*No*	*No*	*Yes*	*No*	1
Field Service (external)	*No*	*No*	*No*	*No*	0
Service Manager	*No*	*Yes*	*No*	*Yes*	2

setting of the modelling session also influences the outcome, the guided interviews also included an evaluation of the modelling session. For example, the participants have been questioned with regard to their perception of the organization of the session and the remote setting due to COVID-19 limitations. The following sections describe the study setting, design and results.

4.1   Case Study Setting

This section specifies and delimits the case investigated in the case study we performed, as recommended by Baxter et al. [4] and Zucker [17]. We investigated the Value Stream for ordering a notebook for employees at the company *Drägerwerk AG & Co. KGaA*, an enterprise located in Lübeck (Schleswig-Holstein, Germany) whose core business model is the development and maintenance of technical, medical and safety equipment all over the world[5]. This Value Stream was already modelled at Dräger in a spreadsheet depicting the main steps, stakeholders, IT components, and known issues. However, we did not investigate the complete Value Stream and left out the part performed at the external hardware vendor due to limited resources.

The investigated Value Stream involves six different groups of stakeholders: the user (i.e., the employee requiring a new notebook for work), the global service desk (assisting the user with purchasing and answering questions that might occur), the team leader (approving any orders of its belonging employees), the client service team (responsible for the procurement and delivery of notebooks for work) and the external field service (executing the delivery of the notebook ordered). Further, we also considered the stakeholder group of the service managers responsible for this Value Stream.

However, we could not cover all six groups of stakeholders with the participants we could convince to participate. In total, we could encourage four participants who are all Dräger employees. Thus, all participants were able to embody the stakeholder group of the user, whereas the stakeholder group of the external field service was not represented at all. Still, no participant was working at the global service desk. Thus, this particular group of stakeholders could not

---

[5] www.draeger.com.

be covered as well. However, one participant was a leader of a team, allowing him to embody that particular stakeholder group. Further, one participant was part of the client service team, whereas two participants were service managers. Table 1 provides an overview which of the participants (A, B, C, And D) was embodying which groups of stakeholders involved.

4.2 Case Study Design

According to Döring et al. [5], a guided interview is a qualitative approach that can provide promising information for yet unexplored and, thus, unpredictable outcomes. Thus, we primarily focused on assessing information from conducting such guided interviews and therefore followed the suggested steps by Döring et al. [5] for a proper execution. Firstly, they state to perform a contextual preparation by determining the topic, the Research Questions, the target group (i.e., the participants and the interviewers), the interview technique (e.g., structured, unstructured, group, single, etc.) and defining as well as practising the guided interview questions. To provide the participants with the knowledge to answer the questions of the guided interview properly, we further carried out a modelling session with the participants roughly following the *its*VALUE framework (see Fig. 1) and using the tool prototype in ADOxx. The participants' perceptions in that modelling session were the topic investigated in the guided interviews afterwards.

As mentioned earlier, the guided interviews have been divided into four parts, each assessing a different method artefact or external factors connected with the modelling session. In this paper, we just focus on the method framework and its concepts. Based on the impressions of the participants, the main question of interest was whether the method fulfils its purpose. This has been detailed into numerous questions addressing different aspects of value-oriented IT-Service modelling as intended by ITIL 4:

RQ 1: Does the method meet its purpose and complies with ITIL 4?
  RQ 1.1: Does it support understanding and explicitly defining Stakeholder Values?
    RQ 1.1.1: How is the VPM perceived regardingly?
    RQ 1.1.2: How is the SVM perceived regardingly?
  RQ 1.2: Does it support understanding and explicitly defining a CVS?
    RQ 1.2.1: How is the VSB perceived regardingly?
  RQ 1.3: Does it support understanding, how Stakeholder Values are affected by several components of a Value Stream?
    RQ 1.3.1: How is the VPM perceived regardingly?
    RQ 1.3.2: How is the VSB perceived regardingly?
  RQ 1.4: Does it support understanding, what actually generates waste?
    RQ 1.4.1: How is the VSB perceived regardingly?
  RQ 1.5: Does it support detecting, where which components (e.g., IT-Systems) are used and information are required or generated?
    RQ 1.5.1: How is the VSB perceived regardingly?

RQ 1.6: Does it support explicitly defining development actions to create an enhanced FVS out of an already existing CVS?

RQ 1.6.1: How is the VPM perceived regardingly?

RQ 1.6.2: How is the SVM perceived regardingly?

RQ 1.6.3: How is the VSB perceived regardingly?

RQ 1.7: Does it support explicitly defining a Value Stream for an entirely new service immediately supporting SV as best as possible?

RQ 1.7.1: How is the VPM perceived regardingly?

RQ 1.7.2: How is the SVM perceived regardingly?

RQ 1.7.3: How is the VSB perceived regardingly?

These questions have been used later as guidelines for the guided interviews after the modelling session. To provide the participants with a good understanding of what *its*VALUE is and how it should work, we performed a participatory modelling session with the participants. However, we had to execute it remotely via Microsoft Teams due to the COVID-19 pandemic. Further, ADOxx does not support any usage by multiple users in parallel. Thus, we had to split up the tasks in the modelling session. Whereas the participants were just requested to discuss and share their input, a host was guiding the discussion and progress of the modelling session. Additionally, the host was also modelling the input the participants gave them. One of the authors of this work embodied both the guiding host and the modelling tool operator during the modelling session due to his unique level of expertise on *its*VALUE. To provide the participants with the current state of the models at any time, the screen of the host was shared for the entire duration of the modelling session. We derived this approach from Sandkuhl et al. [15]'s Sect. 5.1.2 (using a video projector, a guiding host and a modelling tool operator to assist the participants) and 5.2 (using IT-based tools to support a modelling session). Zucker [17] also recommends using further technical tools to carry out and analyze a case study. Although we have not directly analyzed the modelling session itself, we have recorded it to allow checking our perceptions and to provide an evaluation for possible future work.

Due to limited resources and availability of the participants, we had to set the duration for the modelling session to 2:30 h. Consequently, we were just able to perform a single iteration of *its*VALUE on an already existing and rather small Value Stream. Moreover, we requested the participants to provide some preparations in advance to the modelling session. We asked them to take down short notes on what they are actually valuing and what is disturbing or annoying them in the Value Stream for ordering a new notebook for work. Based on their input, we prepared first drafts of all regarding VPMs and a corresponding SVM. Further, we had also prepared a first draft of a VSB we derived from an already existing model for the Value Stream for ordering a new notebook. We also provided all participants with a short introduction (two pages) to our topic in advance to the modelling session, allowing them to receive a general understanding on *its*VALUE. Thereby, we intended to accelerate the first phases of *its*VALUE in order to shift the focus primarily to the detection of waste and definition of development actions. However, we still followed the framework of *its*VALUE (see Fig. 1) by discussing and re-adjusting each prepared VPMs and the SVM followed by the VSB. Afterwards, we focused on collaboratively developing a first draft on how an enhanced (just considering the Stakeholder Values by the participants) Value Stream for ordering a new notebook could look like. As we just investigated the case of enhancing an already existing Value Stream, we highlighted RQ 1.7 due to the regarding perceptions by the participants thus being not as expressive as for the other RQs.

4.3   Evaluation Results

Generally, all participants agreed with the order of the phases we performed during the modelling session. As we stuck to the method framework from Fig. 1 (for the path of improving an already existing Value Stream), this actual order seems to be justified by all participants. Moreover, the participants definitely felt supported by the method in understanding and explicitly defining what different kind of stakeholders value on the Value Stream. Further (with the exception of participant C), they felt also supported in understanding and explicitly defining what the CVS was all about. Additionally, they felt supported in understanding and defining how Stakeholder Values are affected by several components of the Value Stream. Moreover, they felt supported in understanding what actually generates waste inside a Value Stream and what kind of different types of waste exist. Furthermore, they also felt supported in detecting and understanding where which further components of a Value Stream (e.g., IT-Systems or Information) interface (i.e., are used, required, generated, etc.) with other components or processes that are part of the Value Stream. In addition, they also felt supported in identifying and explicitly defining needs for change and development actions in order to achieve a draft for a potentially improved future version of the Value Stream investigated. Although we did not cover the case for using *its*VALUE for designing an entirely new service and thus Value Stream from scratch, all participants (with the exception of C, who felt not capable of providing a proper answer to this question) believe that *its*VALUE is probably suited well correspondingly. Of course, this aspect has to be checked separately on a specifically suiting case. Overall, all different model types seem to suit their purpose in the method of *its*VALUE well. To sum up, the results to the RQs that we defined indicate that *its*VALUE meets its purpose including compliance with the requirements of ITIL 4.

4.4   Limitations

The method perceptions of the method framework and concepts may be influenced by the visual notation, the *its*VALUE modeller (the ADOxx tool prototype), and the circumstances of the modelling session. However, replies from the participants indicate that the influence from the execution of the modelling session and the ADOxx prototype might be relatively small. For the notation, we observed that it (more precisely, the VSB) just had a minor influence on participant C's perception on understanding and defining a Value Stream. However, C later stated (though in the context of identifying needs for change and defining development action) that the actual mechanics of the method are easily understandable, even for people that are not that into modelling and notations. Thus, we conclude that the results regarding the method of *its*VALUE have a satisfying quality, as they probably lead at least in the correct direction and thus deliver good first indications. However, these results are still limited, as we just performed a small proportion of a single iteration of *its*VALUE during the modelling session instead of multiple complete ones as initially recommended. Still, we still considered at least a few examples for any activity that is part of that corresponding path of *its*VALUE during the modelling session. Moreover, the available time for both the modelling session and the guided interviews was limited and thus short. As some participants stated that more time would be required to cover all stakeholders and aspects of a Value Stream (which we actually did not achieve), this should be carefully considered in future work (especially when performing several iterations of *its*VALUE). At last, we were not able to collect data from all involved stakeholder groups.

From an external perspective, we have just investigated one single case. Consequently, a generalization from that particular single case to several other ones that we have not investigated yet accordingly is not reasonably possible according to Zucker [17] and Gerring [7]. Thus, our results and conclusions are just applicable to our specific case investigated. To reasonably draw a generalization, additional case studies should be performed in different scenarios with much more participants or even other evaluation techniques (e.g., questionnaires). Further, we have just gathered findings for *its*VALUE on the *perceived* efficacy and intention to use, whereas information on the actual efficacy and usage still remain unknown. This has to be investigated in future work as well.

5   Conclusion and Outlook

*its*VALUE in combination with the modeller supports comprehensive modelling and analysis of IT-Service related Value Streams. Based on its consideration of ITIL 4 concepts, it has the potential to support practitioners in adopting that standard. A first case study on a hardware purchasing service for the evaluation of *its*VALUE showed a great relevance of models and analysis results for value-oriented IT-Service modelling according to the involved stakeholders. Furthermore, a majority of case study participants showed interest in the future use of *its*VALUE method and modeller. Considering the limitations of the discussed case study, a further evaluation is required. Though there are case study results regarding the visual notation and the modeller that are not deeply discussed in this paper, these results also indicate the appropriateness of the method for its purpose. Still, these additional artefacts need further evaluation. The visual notation was not explicitly investigated for each model type of *its*VALUE separately. However, their quality seems to differ according to the answers of the participants. Thus, their statements regarding the notations for each model type are mainly limited to their general overall perception. In future work, each model type should be investigated separately regarding its notation quality. Additionally, as the participants did not actually use the modeller (ADOxx tool prototype) on their own, their perception and thus our results on it are also limited. Thus, actual usage experiences must be gathered in future work.

The feedback of this case study and future evaluations will help to better adjust and refine the approach for practitioners. Having the modeller freely available at OMiLAB[6] assures access for and involvement of potential users. An important next step will be the development of a method guideline that fits the needs of practitioners. Furthermore, complexity handling needs to be evaluated in a more complex scenario compared to the relatively simple scenario of our first case study.

References

1. AXELOS: ITIL 4 Managing Professional Create, Deliver and Support. The Stationery Office (2019)
2. AXELOS: ITIL 4 Managing Professional Drive Stakeholder Value. The Stationery Office (2019)
3. AXELOS: ITIL FOUNDATION, ITIL 4 Edition (GERMAN EDITION). The Stationery Office (2019)
4. Baxter, P., Jack, S., et al.: Qualitative case study methodology: study design and implementation for novice researchers. Qual. Rep. **13**(4), 544–559 (2008)
5. Döring, N., Bortz, J.: Forschungsmethoden und evaluation. Springer Verlag, Wiesbaden (2016). https://doi.org/10.1007/978-3-642-41089-5
6. Geist, M.R.: Using the Delphi method to engage stakeholders: a comparison of two studies. Eval. Program Plan. **33**(2), 147–154 (2010)
7. Gerring, J.: What is a case study and what is it good for? Am. Polit. Sci. Rev. **98**(2), 341–354 (2004)
8. Hartmann, L., Meudt, T., Seifermann, S., Metternich, J.: Value stream design 4.0: designing lean value streams in times of digitalization and Industrie 4.0 [wertstromdesign 4.0: Gestaltung schlanker wertstroeme im zeitalter von digitalisierung und Industrie 4.0]. ZWF Zeitschrift fuer Wirtschaftlichen Fabrikbetrieb **113**(6), 393–397 (2018)
9. Hartmann, L., Meudt, T., Seifermann, S., Metternich, J.: Value stream method 4.0: holistic method to analyse and design value streams in the digital age. Procedia CIRP **78**, 249–254 (2018)
10. Heger, S., Valett, L., Thim, H., Schröder, J., Gimpel, H.: Value stream model and notation-digitale transformation von wertströmen. In: Wirtschaftsinformatik (Zentrale Tracks), pp. 710–724 (2020)
11. Meudt, T., Leipoldt, C., Metternich, J.: Der neue blick auf verschwendungen im kontext von Industrie 4.0: Detaillierte analyse von verschwendungen in informationslogistikprozessen. Zeitschrift für wirtschaftlichen Fabrikbetrieb **111**(11), 754–758 (2016)
12. Meudt, T., Metternich, J., Abele, E.: Value stream mapping 4.0: holistic examination of value stream and information logistics in production. CIRP Ann. **66**(1), 413–416 (2017)

---

[6] https://www.omilab.org/.

13. Nerdinger, F.W.: Teamarbeit. In: Arbeits- und Organisationspsychologie. Springer-Lehrbuch, pp. 103–118. Springer, Heidelberg (2014). https://doi.org/10.1007/978-3-642-41130-4_8
14. Richter, Henning snd Lantow, B.: It-service value modeling: a systematic literature analysis. In: 12th Workshop on Business and IT Alignment (BITA 2021)). Accepted Paper (2021)
15. Sandkuhl, K., Wißotzki, M., Stirna, J.: Begriffe im Umfeld der Unternehmensmodellierung, pp. 25–40. Springer, Heidelberg (2013). https://doi.org/10.1007/978-3-642-31093-5_3
16. Schoenwaelder, M., Szilagyi, T., Baer, F., Lantow, B., Sandkuhl, K.: It self-service blueprinting a visual notation for designing it self-services. In: Bork, D., Lantow, B.G.J. (eds.) CEUR Workshop Proceedings, vol. 2238, pp. 88–99. CEUR-WS (2018)
17. Zucker, D.M.: How to do case study research. School of Nursing Faculty Publication Series, p. 2 (2009)

Architecting Intelligent Service Ecosystems: Perspectives, Frameworks, and Practices

Alfred Zimmermann[1]([⊠]) [iD], Rainer Schmidt[2] [iD], Kurt Sandkuhl[3] [iD],
Yoshimasa Masuda[4,5] [iD], and Abdellah Chehri[6] [iD]

[1] Reutlingen University, Reutlingen, Germany
`alfred.zimmermann@reutlingen-university.de`
[2] Munich University of Applied Sciences, Munich, Germany
`rainer.schmidt@hm.edu`
[3] University of Rostock, Rostock, Germany
`kurt.sandkuhl@uni-rostock.de`
[4] Keio University, Tokyo, Japan
`yoshi_masuda@keio.jp`, `ymasuda@andrew.cmu.edu`
[5] Carnegie Mellon University, Pittsburgh, USA
[6] University of Quebec at Chicoutimi, Saguenay, Canada
`abdellah_chehri@uqac.ca`

Abstract. The current advancement of Artificial Intelligence (AI) combined with other digitalization efforts significantly impacts service ecosystems. Artificial intelligence has a substantial impact on new opportunities for the co-creation of value and the development of intelligent service ecosystems. Motivated by experiences and observations from digitalization projects, this paper presents new methodological perspectives and experiences from academia and practice on architecting intelligent service ecosystems and explores the impact of artificial intelligence through real cases supporting an ongoing validation. Digital enterprise architecture models serve as an integral representation of business, information, and technological perspectives of intelligent service-based enterprise systems to support management and development. This paper focuses on architectural models for intelligent service ecosystems, showing the fundamental business mechanism of AI-based value co-creation, the corresponding digital architecture, and management models. The focus of this paper presents the key architectural model perspectives for the development of intelligent service ecosystems.

Keywords: Service ecosystems · Value co-creation · Intelligent digital architecture · Architecture engineering · Management

1 Introduction

Intelligent service ecosystems together with their digital platforms [1] are now considered one of the most important foundations for new business models that contribute to developing and implementing corporate strategies. In 2019, seven of the ten most valuable companies in the world provided digital platforms for service ecosystems (e.g.,

© Springer Nature Switzerland AG 2021
R. A. Buchmann et al. (Eds.): BIR 2021, LNBIP 430, pp. 150–164, 2021.
https://doi.org/10.1007/978-3-030-87205-2_10

Microsoft, Facebook, Alibaba, Amazon) [1]. Together with their service ecosystems, platforms generate value by inducing collaboration and co-creation between actors and facilitating actions between external consumers and producers of products and services [2]. Artificial intelligence plays a crucial role in the design of such intelligent service ecosystems [1].

Crucial to the design of intelligent service ecosystems are creative ideas, competencies, and capabilities for intelligent digitization, with timely adaptation to ever shorter innovation cycles with appropriate implementation and execution concepts within suitable business models. However, there is a lack of substantial methodological experience from academia and practice, especially case studies that take a holistic perspective on designing intelligent service ecosystems.

The essential ultimate success of a platform does not result primarily from the platform itself or the technology's use. Instead, successful platforms result from the so-called ecosystem of the platform. The ecosystem of the platform is based on the totality of users who collaborate via the platform. These users can be distinguished into several groups. To establish a platform, it is necessary to achieve sufficient growth in several groups. Network effects play a central role in this process. They can occur within but also between user groups. Fostering these network effects is a critical task in the development of a platform. An ecosystem is characterized by a set of interacting resources, such as organizations, individuals, and autonomous actors and systems, that jointly develop their capabilities and roles according to the Service-Dominant Logic (S-D logic) [3]. From a technical perspective, digital services are slices of code that perform a specific functionality to enable business offerings [4] of digital products and services composed of business services, data services, and infrastructure services.

Based on the S-D logic [5], a service ecosystem is a self-contained and self-adaptive system of loosely coupled resource-integrating actors connected by shared institutional logic and mutual value creation through service exchange. S-D logic was initially developed in 2004 by Stephen Vargo and Robert Lusch [6]. S-D logic represents a new type of service logic that focuses on intangible resources, the co-creation of value, and close business relationships. S-D logic focuses on an emerging understanding of economic exchange that favors services based on specialized skills over traditional manufactured goods.

The following *research questions* can substantiate our approach:

RQ1: How does artificial intelligence impact value co-creation in service ecosystems using a Service-Dominant Logic theoretical lens?
RQ2: What are integral perspectives for architecting and managing intelligent service ecosystems?

We will present new perspectives and frameworks from academia and practice for architecting AI-based service ecosystems. First, we introduce the research background and methodology base. The paper's core addresses our research questions: (1) we present AI-powered co-creation mechanisms, (2) we provide new perspectives motivated by AI and associated digital technologies in our comprehensive and holistic approach to architecting and managing intelligent service ecosystems. Finally, we summarize our discussion from science and conclude our findings and future research.

2 Research Background

We are at a turning point in the development and application of intelligent digital systems. We see excellent prospects for digital systems with artificial intelligence (AI) [7, 8] to contribute to improvements in many areas of work and society with the potential of digital technologies. We understand digitalization based on new methods and technologies of AI as a complex integration of digital services, products, and associated systems, with a high degree of automation, autonomy, and self-adaptation. Artificial intelligence is receiving a high level of attention due to recent advances in various areas such as image recognition, translation, and decision support [9]. Advances in AI are driving the changing role of technology for service ecosystems. The contribution of artificial intelligence can be categorized in terms of the four roles of technology [10]: assistive (human-in-the-loop, hard-wired system), augmentative (human-in-the-loop, adaptive system), automating (no-human-in-the-loop, hard-wired system), and autonomous (no-human-in-the-loop, adaptive system). The assistive type of AI technology [2] fits into the traditional domain of service ecosystems and service science [11] of using technology as a strategic driver and improving service functionality. However, the remaining three categories, augmenting, automating, and autonomous AI technologies, imply an increasing degree of agency and direct interactions with humans and the environment. The changing role of technology, from a tool to an "actor" in value co-creation [12], requires a new conceptualization of technology in service science [13]. Service system [14] entities are responsible actors, as in [11], and clarifies under what conditions a technology such as a cognitive assistant (CA) becomes a responsible actor.

Data, computation currently drive artificial intelligence progress and advances in machine learning, perception and cognition, planning, and natural language algorithms. AI enables exciting new business applications such as predictive maintenance, logistics optimization, and customer service management improvement. Artificial intelligence supports decision-making in many business domains. Therefore, most companies expect AI to provide a competitive advantage. Today's advances in AI [15, 16] have led to a rapidly growing number of intelligent services and applications. The collaborative development of capabilities through intelligent digital systems promises significant benefits for science, business, and society.

In contrast to symbolic AI [7], machine learning [17] uses an inductive approach based on a large amount of analyzed data. We distinguish three basic machine learning approaches [15, 16]: supervised, unsupervised, and reinforcement learning. In supervised machine learning approaches, the target value is part of the training data and is based on sample inputs. Typically, unsupervised learning is used to discover new hidden patterns within the analyzed data. Reinforcement learning (RL) is an area of machine learning with software agents [8] that maximizes cumulative rewards. The exploration environment is specified in terms of a Markov process, as many reinforcement learning algorithms use dynamic programming techniques. Reinforcement learning does not require labeled input/output pairs, and suboptimal actions do not need to be explicitly corrected.

Artificial intelligence is often characterized as impersonal: From this perspective, intelligent systems operate automatically and independently of human intervention. The

public discourse on autonomous algorithms working on passively collected data contributes to this view. However, this prospect of huge automation obscures the extent to
which human work necessarily forms the basis for modern AI systems and makes them
possible in the first place. The human element of intelligent systems includes optimizing knowledge representations, developing algorithms, collecting and tagging data, and
deciding what to model and interpret the results. The study of artificial intelligence [18]
from a human-centric perspective requires a deep understanding of the role of human
ethics, human values and customs, and the practices and preferences for development and
interaction with intelligent systems. With the success of AI, new concerns and challenges
regarding the impact and risks of these technologies on human life are emerging. These
include issues of the limited feasibility of AI-based systems, the security and trustwor-
thiness of AI technologies in digital systems, the fairness and transparency of systems,
the still limited explainability of derived solutions and decisions, and the intended and
unintended impacts of AI on people and society.

Enterprise Architecture (EA) [19, 20] has evolved for more than a decade as a
discipline with a scientific background and functional decision-support capabilities and
models for forward-looking AI technology-based businesses and digital organizations
[21]. Enterprise architecture aims to model, align, and understand essential interactions
between business and IT to set the stage for a well-aligned and strategically oriented
decision framework for both digital business and digital technologies [22].

3 Value Co-creation in Platforms and Ecosystems

Platform strategies are becoming a critical component for many business models [23].
Platforms enable direct interactions between multiple otherwise unconnected groups of
actors [24]. Platforms enable interactions between two or more groups of actors. A shift
in strategic focus to building communities and engaging resources of platform members
is identified as a new perspective in [25].

During the evolution of platforms, different ways of value creation were used [26].
Product platforms such as Windows or IBM 360 used complementary products to create value. On social platforms the creation of network effects and ecosystems through
platforms provides a new source of competitive advantage [27].

So far, a strict separation between value producer and value consumer was the basis
of value creation models. In recent years, however, there has been a significant change.
The focus of research is now on value co-creation models [28], in which value is created
through the interaction of several partners [6]. Typically, the former consumer takes an
increasingly active role, which leads to his role being referred to as prosumer [29].

Lusch and Nambisan [12] developed a model for value co-creation from a service innovation perspective using an SD-Logic [3] theoretic lens. The basis for service
provisioning is resource liquefaction, in which the resources are made accessible and
manageable regardless of their physical presence (4). The aim of the whole is to increase
resource density, i.e., the accuracy of fit of the services offered and requested (5). Platforms foster resource liquefaction and resource density. In [2], the model has been used
to develop a model for value co-creation on platforms, Fig. 1. The platform in the middle
of the value co-creation model links different groups of actors. The following phases

are differentiated: In the first phase, there is an exchange of information about the value proposition (1). The platforms enhance the exchange of information on service propositions and thus increases resource liquefaction. In the second phase, the suggestions are filtered so that the suggestions of interest to the respective actuators are selected (2). The filtering improves the fit of service propositions and thus the resource density. Finally, in the third phase, the exchange of services takes place. For this purpose, the operational execution of the service exchange is to be ensured (3).

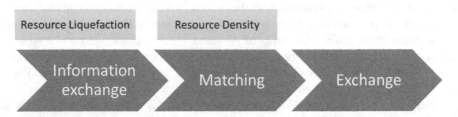

Fig. 1. SD-logic-based value co-creation model for platform ecosystems [2].

Artificial intelligence can be used in all three phases of the above model. Artificial intelligence-based interfaces can facilitate the exchange of information to actors, such as is done on assistant-based platforms [2]. The filtering of offers and requests based on the similarity function is a classic application of artificial intelligence. The automation capabilities of artificial intelligence come into play in the third phase of the model, the exchange of services. Artificial intelligence coordinates the actions of the actors. For example, when renting an apartment on Airbnb [30], granting access to the apartment and releasing the apartment are physical operations that need to be coordinated.

Artificial intelligence improves resource liquefaction, e.g., by improving access to information on hitherto unstructured data sources that describe the resource. For example, artificial intelligence can extract the powerful feature from a service description and tag it appropriately. In this way, artificial intelligence decouples information from its physical form or device [12]. Resource density is the principle of mobilizing contextual information on platforms as effectively and efficiently as possible [12]. Artificial intelligence increases it by improving the coordination of service offers and inquiries.

4   Intelligent Service Architecture

According to [31], a service ecosystem is a self-contained, self-adjusting system of loosely coupled recourse integrating actors connected by value co-creation through service exchange. In our understanding, a successful digital service platform [32] should support a network of actors and host a set of loosely coupled open services and software products as part of a rapidly growing digital ecosystem [12]. The DEA Cube (Digital Enterprise Architecture Reference Cube) in Fig. 2 extends our holistic architecture reference and classification framework from [21] to drive bottom-up integration of dynamically composed micro-granular architecture services and their models. Furthermore, the DEA Cube abstracts from a particular business scenario or technology

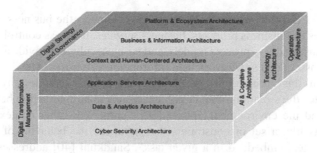

Fig. 2. Digital enterprise architecture reference cube (DEA Cube).

because it can be applied to different architectural instantiations to support intelligent ecosystems independently of different domains.

Metamodels and their architectural data [33] are the core part of a digitalization architecture. Architecture metamodels should support analytics-based architectural decision management and the strategic as well as IT/business alignment. Three quality perspectives are essential for an adequate IT/business alignment and are differentiated as (I) IT system qualities: performance, interoperability, availability, usability, accuracy, maintainability, and suitability; (II) business qualities: flexibility, efficiency, effectiveness, integration and coordination, decision support, control and follow up, and organizational culture; and finally (III) governance qualities: plan and organize, acquire and implement deliver and support, monitor and evaluate.

DEA addresses first the top of the Platform and Ecosystem Architecture [12, 32]. A digital platform is a repository of business, data, and infrastructure services used to configure digital offerings from digital services rapidly. Digital services and components are slices of code that perform a specific task. We position reusable digital services as parts of an ecosystem of services. Further, a digital platform linearizes the complexity of cooperating services. The value of a platform to users [32] results from the number of platform and service users. Platforms do not own or control their resources and are therefore well suited for scalability within the ecosystem. DEA extends the platform and ecosystem architecture by a set of close related architectural viewpoints.

A digital strategy [34] is a combination of initiatives where a company will select online activities to help realize its digital business objectives/vision. Digital governance [35] should additionally set the frame for digital strategies and digital innovation management. The second aim of governance is to define and assess rules for value-oriented digital compliance.

Five strategic domains define the focus of Digital Transformation Management, as in [36]: customers, competition, data, innovation, and value. Customers' most important strategical changes in a digital business are: customers as a dynamic network, two-way communication for co-creation, key influencers, marketing to inspire purchase and loyalty, common value flows, and economies of value. Strategic changes also affect competitors, as competition across fluid industries, blurred differentiation between partners and competitors, competitors cooperate in critical areas (coopetition). Key assets reside in external networks, platforms, and ecosystems with partners that exchange value and all gain due to network effects.

The Business & Information Architecture [37] connects the business strategy with model structures for business products, business services, business control information, business domains, business process models, and business rules to provide a specification framework for associated service-oriented information systems.

According to the basic definition of Dey [38], a context includes any information that characterizes the situation of an entity and relates to an interaction between users, applications, and the environment. Bazire [39] summarizes other context definitions as "context acts like a set of constraints that influence the behavior of a system (a user or a computer) embedded in a given task". Sandkuhl [40] addresses an original context modeling approach for enterprise IT applications to support human actors. An information demand context model is an extract from an enterprise model for a specific role, taking into account all the roles' tasks and linked to the specific resources.

The AI and Cognitive Architecture outlines fundamental components, services, mechanisms, and methods to support the intelligent behavior of evolved digital systems and services. Artificial intelligence and cognitive computing simulate and extend human thinking and intend to mimic the working model of the brain [41] without knowing exactly how the biology of the brain works. Cognitive computation [42] is inspired by neurobiology, cognitive psychology, AI, and connectionism. Connectionism represents a cognitive theory executed by adaptive and learning neural networks. IBM's cognitive computing program [43] focuses on next-generation information systems that leverage AI technologies and data analytics to enable understanding, reasoning, learning, and interaction. Cognitive systems continuously build knowledge through learning, understand natural language, support problem solving, and interact with humans more naturally than traditional programmed systems. A significant aspect of AI technology is robust or trustworthy AI. Transparency includes ways to explain AI results (e.g., why a system recommends a particular course of action), metrics to measure the effectiveness of an AI algorithm, verification and validation of intelligent systems, computer security, and the regulatory aspects that govern the safe, responsible, and ethical use of AI technology.

The Data Architecture [44] describes and classifies the data structures used by an enterprise and its computer application software. A data architecture consists of models, policies, rules, or standards that govern what data is collected and how it is stored, arranged, integrated, and used in data systems and organizations. Data architectures deal with data in storage, data in use, and data in motion; descriptions of data stores, data groups, and data elements; and mappings of these data artifacts to data qualities, applications, locations, etc.

The Application Services Architecture [4, 45] is the software reference architecture of application services that compiles the main application-specific service types and defines their relationship through a layered model by services that build on each other. The core functionality of domain services is linked with application interaction services and with business processes services of the customer organization. The core functionality of domain services is linked with application interaction capabilities and with the business processes of the customer's organization.

The Cybersecurity Architecture [46] specifies the organizational structure, standards, policies, and functional behavior of a computer network, including security and network functions. Cybersecurity architecture is also how the various components of a cyber or computer system are organized, synchronized, and integrated. A cybersecurity architecture framework is a component of the overall architecture of a system and guides the design of an entire product/system.

The Technology Architecture [47] models domain-agnostic software and hardware platforms to support the deployment of business, data, and application services. Technology includes IT infrastructures, platforms, middleware, networks, communications, processing, and related standards.

The Operations Architecture [48] relies on service management processes to enable the ongoing support and management of AI-based infrastructures of a digital enterprise. A company's IT infrastructure typically consists of many different systems and platforms, often located in different geographic locations. Kubernetes is an open-source container orchestration system initially developed by Google to automate the deployment, scaling, and management of service-based applications. Kubernetes [49] provides a platform for automating the deployment, scaling, and operation of application containers for images created with Docker [50] across clusters of hosts.

5   AIDAF Framework for Digital Platforms and Service Ecosystems

We have shown in [51] that companies that have previously applied TOGAF [19] or FEAF can successfully use the AIDAF integrated EA framework to support cloud computing when strategically promoting cloud/mobile IT. The model proposed by AIDAF [52] with the Architecture Board (AB) is shown in Fig. 3, as below. The AIDAF is an EA framework integrating an adaptive EA cycle for different business units. It involves the Architecture Board performing architecture reviews and enabling the alignment between IT architecture strategy and solution architecture in information system projects, including digital IT solutions [51, 52]. Therefore, the AIDAF framework is essential and necessary for developing a digital IT strategy and supporting digital transformation.

In the adaptive EA cycle, IS/IT project plan documents with the architecture of new digital IT projects can be developed on a short-term basis initially. Context phase refers to defining phase materials, where architectural guidelines for Cloud services, security, and digital IT can be defined as common ones in the enterprise, per business needs and demands. During the Assessment/Architecture Review phase, the AB should direct and review the architecture in the IS/IT project [53].

In the Rationalization phase, the stakeholders and AB decide upon replaced or decommissioned systems by the proposed new information systems. The equivalent project team can begin implementing the new digital IT project after deliberating issues and action items [51–53]. In the adaptive EA cycle, corporations can adopt an EA framework like TOGAF and simple EA framework that is a simple mid- to long-term perspective EA structure composed of Deliverables (target architecture, roadmaps), EA processes (financial/budgeting approval process by deliverables) and Principles (architecture directions, policy), based on an operational division unit in the upper part of the above Fig. 3, equivalent to Figure 1 of [51–53] in alignment between EA guiding principles and each division's principles, corresponding to differing strategies in business divisions in the mid-long-term. TOGAF Architecture Development Method (ADM)

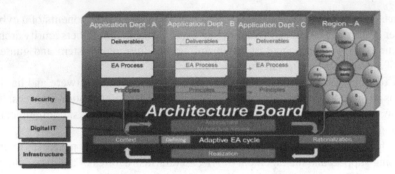

Fig. 3. AIDAF model with architecture board and governance.

describes a step-by-step approach to developing enterprise architecture as a core element [53].

We refer to cases of applying the AIDAF framework to Industry 4.0 Architecture – RAMI 4.0 [54]. At the enterprise level, we show a digital platform usage scenario for a drug platform board (DPB) in Fig. 4.

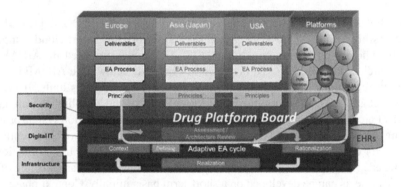

Fig. 4. AIDAF model for DPB.

In particular, the Architecture Board, the example of the EA framework structure in a specific global healthcare company examined in previous papers [51, 54]. In a global EA rollout, we are handling cloud, mobile IT, and big data strategic projects and systems that took priority in Europe and US Group companies well by structuring and implementing EA with the above AIDAF to be consistent with global IT strategy focusing on cloud, mobile IT, big data and digital IT [51, 53].

The AIDAF model with DPB covers the core drug digital platforms, such as drug development and clinical decision support (CDS) platforms, that are used by new drug development planners, managers, digital IT practitioners in each region and are reviewed by the DPB. We describe the CDS digital platform [55] in Fig. 5 at an ecosystem level.

The CDS platform provides recommendations based on the available patient-specific data (EHR) and medical facts (knowledge base) among the healthcare ecosystem partners [55].

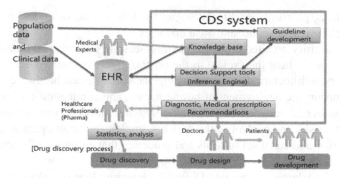

Fig. 5. Reference architecture for CDS system in the ecosystem [55].

6 Discussion

We are summarizing our results and synthesizing our core findings and experiences. A digital platform and an ecosystem should enable shared value creation for all stakeholders and facilitate the exchange of goods, services, and social currency. DEA - The Digital Enterprise Architecture Reference Cube provides our new architectural reference model for the bottom-up integration of dynamically composed micro-granular architecture services and their models. A DEA-aligned platform integrates core technology services as a base that provide standardized access points and repositories for intelligent service ecosystems composed of business services, data services, and infrastructure services. Artificial intelligence is a powerful and useful technology to support essential functionalities of intelligent service ecosystems and the Service-Dominant Logic-grounded value co-creation process.

The newly introduced architecture domain in DEA of the Context- and Human-Centered Architecture, together with the realigned AI & Cognition Architecture, supports the growing set of automated functionalities that do not follow an explicit preformulated model. The DEA Cube provides a comprehensive architectural reference model to compose micro-granular architecture service models, like the Internet of Things and Microservices, to support intelligent digital services and products.

The DEA perspective of digital strategy [34, 56, 57] and governance [35] defines the base for well-aligned management practices through specifying essential architectural management activities: plan, define, enable, measure, and control. The business & information architecture establishes the link between the enterprise business strategy and the results of supporting strategic initiatives through information systems.

The context and human-centric architecture ensure the treatment of system-wide contexts and user-specific models to enable dynamic adaptability and customizability of system behavior. AI models various mental processes using computers, while cognitive computing simulates human brain functions as a computational model. The canonical architecture of an AI system [58] outlines key components to support the development of an effective AI solution. The canonical AI architecture includes sensors and other data sources and components for data conditioning, AI algorithms, human-machine teaming, user capabilities, processing technologies, and robust AI. The data analytics architecture refers to systems, protocols, and technologies used to collect, store, and analyze data based on machine learning and statistical analytics.

The cybersecurity architecture helps position security controls and breach counter-measures and shows how they relate to the organization's overall system framework. The technology architecture describes fundamental software and hardware capabilities required to support the deployment of business, data, and application services. Finally, the operations architecture ensures that these systems function correctly by automating operational tasks and control of the systems. Implementing an operations architecture consists of a dedicated set of tools and processes that automate IT operations in a coordinated manner.

The AIDAF (Adaptive Integrated Digital Architecture Framework) is a new digital architecture framework that can address the digital agility elements for digital IT strategy and digital transformation lack in existing traditional EA frameworks [53].

7 Conclusions and Future Research

We identified the needs and solution mechanisms for architecting intelligent service ecosystems by presenting our new AI-induced, augmented, and integral methodological perspectives and frameworks. We first outlined the research background and presented our applied methodology to create and evaluate a new type of enterprise architecture by applying architectural methods from science and practice. Next, we set this transparent methodological background in the context of AI-based digitalization, in which we map the core results to fit our research questions. We have established an integral view by introducing new perspectives arising from the fusion of AI with service ecosystems and digital architectures by adequately aligning comprehensive frameworks and methodologies for architecting intelligent service ecosystems. First, in our architectural approach, we introduced AI-enhanced co-creation mechanisms in line with the Service-Dominant Logic (SD-L) as a prerequisite for intelligent service ecosystems on digital platforms. Second, we presented an advanced service reference architecture for intelligent service ecosystems and an integrated approach to adaptive architecture engineering and management. For this purpose, we aligned the DEA - Digital Enterprise Architecture Reference Cube with AIDAF - the Adaptive Integrated Digital Architecture Framework. Last, we have summarized our findings and recommendations from methodological experiences from science and practice for a new and innovative style of architecture of intelligent service ecosystems.

The strengths of our research result from our novel approach to support intelligent digitalization for architecting intelligent service ecosystems. We have integrated an AI-powered co-creation model with an integral and scalable digital architecture reference model in conjunction with the framework for adaptive architecture development and management. Limitations of our work arise from an ongoing validation of our research and open questions of comprehensive AI approaches and related inconsistencies and semantic dependencies. In particular, we cannot quantify the impact of artificial intelligence or specific technologies on mechanisms and elements in intelligent service ecosystems. Furthermore, we still need to prioritize artificial intelligence technologies for supporting specific tasks in intelligent service ecosystems.

Our future research will investigate in more detail how artificial intelligence affects value creation in service ecosystems, particularly resource liquefaction and resource density. Artificial intelligence is a very diverse technology with different types of model building and problem-solving capabilities. Therefore, an important research task will be to identify those artificial intelligence technologies that are particularly suited to support the creation of intelligent service ecosystems. Future research will also address mechanisms, reference architectures, methodologies, and guidelines for the flexible and adaptive integration of intelligent digital architectures for AI-based service ecosystems based on study results from use cases of digital transformation and our educational experiences from academia and practice.

References

1. McAfee, A., Brynjolfsson, E.: Machine, Platform, Crowd: Harnessing Our Digital Future. W. W. Norton & Company, New York (2017)
2. Schmidt, R., Alt, R., Zimmermann, A.: A conceptual model for assistant platforms. In: 54th Hawaii International Conference on System Sciences (HICSS), Wailea, pp. 4024–4033 (2021)
3. Vargo, S.L., Lusch, R.F.: Institutions and axioms: an extension and update of service-dominant logic. J. Acad. Mark. Sci. 44(1), 5–23 (2015). https://doi.org/10.1007/s11747-015-0456-3
4. Newman, S.: Building Microservices: [Designing Fine-Grained Systems]. O'Reilly, Beijing (2015)
5. Subramaniam, M., Iyer, B., Venkatraman, V.: Competing in digital ecosystems. Bus. Horiz. 62, 83–94 (2019)
6. Vargo, S.L., Lusch, R.F.: Evolving to a new dominant logic for marketing. J. Market. 68, 1–17 (2004)
7. Russell, S.J., Norvig, P.: Artificial Intelligence: A Modern Approach. Pearson Education Limited, Malaysia (2016)
8. Poole, D.L., Mackworth, A.K.: Artificial Intelligence: Foundations of Computational Agents. Cambridge University Press, Cambridge (2017)
9. Brynjolfsson, E., McAfee, A.: The Second Machine Age: Work, Progress, and Prosperity in a Time of Brilliant Technologies. W. W. Norton & Company, New York (2014)
10. Rao, A.S., Verweij, G.: Sizing the prize what's the real value of AI for your business and how can you capitalize? PwC (2017)
11. Siddike, M.A.K., Hidaka, K., Kohda, Y.: Technology as actors in service systems. In: Proceedings of the 54th Hawaii International Conference on System Sciences, pp. 1030–1039 (2021)
12. Lusch, R.F., Nambisan, S.: Service innovation: a service-dominant logic perspective. MIS Q. 39, 155–175 (2015). https://doi.org/10.25300/MISQ/2015/39.1.07
13. Maglio, P., Spohrer, J.: Fundamentals of service science. J. Acad. Mark. Sci. 36, 18–20 (2008). https://doi.org/10.1007/s11747-007-0058-9
14. Maglio, P., Vargo, S., Caswell, N., Spohrer, J.: The service system is the basic abstraction of service science. IseB 7, 395–406 (2009). https://doi.org/10.1007/s10257-008-0105-1
15. Munakata, T.: Fundamentals of the New Artificial Intelligence: Neural, Evolutionary, Fuzzy and More. Springer, London (2008). https://doi.org/10.1007/978-1-84628-839-5
16. Skansi, S.: Introduction to Deep Learning: From Logical Calculus to Artificial Intelligence. Springer, Cham (2018). https://doi.org/10.1007/978-3-319-73004-2
17. Hwang, K.: Cloud Computing for Machine Learning and Cognitive Applications. MIT Press, Cambridge (2017)

18. Kearns, M., Roth, A.: The Ethical Algorithm: The Science of Socially Aware Algorithm Design. Oxford University Press, New York (2019)
19. Lankhorst, M.: Enterprise Architecture at Work. Springer, Heidelberg (2017). https://doi.org/10.1007/978-3-662-53933-0
20. Nurmi, J., Pulkkinen, M., Seppänen, V., Penttinen, K.: Systems approaches in the enterprise architecture field of research: a systematic literature review. In: Aveiro, D., Guizzardi, G., Guerreiro, S., Guédria, W. (eds.) EEWC 2018. LNBIP, vol. 334, pp. 18–38. Springer, Cham (2019). https://doi.org/10.1007/978-3-030-06097-8_2
21. Zimmermann, A., Schmidt, R., Sandkuhl, K., Jugel, D., Schweda, C., Bogner, J.: Architecting digital products and services. In: Zimmermann, A., Schmidt, R., Jain, L.C. (eds.) Architecting the Digital Transformation. ISRL, vol. 188, pp. 181–197. Springer, Cham (2021). https://doi.org/10.1007/978-3-030-49640-1_10
22. Niemi, E., Pekkola, S.: Using enterprise architecture artefacts in an organisation. Enterp. Inf. Syst. 11, 313–338 (2017)
23. Wan, J., et al.: A manufacturing big data solution for active preventive maintenance. IEEE Trans. Industr. Inf. 13, 2039–2047 (2017)
24. Hagiu, A.: Strategic decisions for multisided platforms. MIT Sloan Manage. Rev. 55, 71–80 (2014)
25. Alstyne, M.W.V., Parker, G.G., Choudary, S.P.: Pipelines, platforms, and the new rules of strategy. Harvard Bus. Rev. 94, 54–62 (2016)
26. Cusumano, M.A., Gawer, A., Yoffie, D.B.: The future of platforms. MIT Sloan Manage. Rev. Spec. Issue Disrupt. 61, 46–54 (2020)
27. Parker, G., van Alstyne, M.W., Choudary, S.P.: Platform Revolution: How Networked Markets are Transforming the Economy–and How to Make Them Work for You. Norton & Company, New York (2016)
28. Galvagno, M., Dalli, D.: Theory of value co-creation: a systematic literature review. Manag. Serv. Qual. 24, 643–683 (2014). https://doi.org/10.1108/MSQ-09-2013-0187
29. Klein, S., Totz, C.: Prosumers as service configurators-vision, status and future requirements1. E-Life After the Dot Com Bust. 119 (2004)
30. Camilleri, J., Neuhofer, B.: Value co-creation and co-destruction in the Airbnb sharing economy. IJCHM. 29, 2322–2340 (2017). https://doi.org/10.1108/IJCHM-09-2016-0492
31. Tiwana, A.: Platform Ecosystems: Aligning Architecture, Governance, and Strategy. Morgan Kaufmann, Amsterdam (2013)
32. Ross, J.W., Beath, C.M., Mocker, M.: Designed for Digital: How to Architect Your Business for Sustained Success. MIT Press, New York (2019)
33. Lagerström, R., Johnson, P., Ekstedt, M.: Architecture analysis of enterprise systems modifiability: a metamodel for software change cost estimation. Software Qual J. 18, 437–468 (2010). https://doi.org/10.1007/s11219-010-9100-0
34. Bones, C., Hammersley, J., Shaw, N.: Optimizing Digital Strategy. Kogan Page (2019)
35. Weill, P., Ross, J.W.: IT Governance: How Top Performers Manage IT Decision Rights for Superior Results. Harvard Business School Press, Boston (2004)
36. Rogers, D.L.: The Digital Transformation Playbook: Rethink Your Business for the Digital Age. Columbia University Press, New York (2016). https://doi.org/10.7312/roge17544
37. Řepa, V., Svatoš, O.: Adaptive and resilient business architecture for the digital age. In: Zimmermann, A., Schmidt, R., Jain, L.C. (eds.) Architecting the Digital Transformation. ISRL, vol. 188, pp. 199–221. Springer, Cham (2021). https://doi.org/10.1007/978-3-030-49640-1_11
38. Dey, A.K.: Understanding and using context. Pers. Ubiquit. Comput. 5, 4–7 (2001). https://doi.org/10.1007/s007790170019

39. Bazire, M., Brézillon, P.: Understanding context before using it. In: Dey, A., Kokinov, B., Leake, D., Turner, R. (eds.) CONTEXT 2005. LNCS (LNAI), vol. 3554, pp. 29–40. Springer, Heidelberg (2005). https://doi.org/10.1007/11508373_3
40. Sandkuhl, K., Borchardt, U.: How to identify the relevant elements of "context" in context-aware information systems? In: Johansson, B., Andersson, B., Holmberg, N. (eds.) BIR 2014. LNBIP, vol. 194, pp. 290–305. Springer, Cham (2014). https://doi.org/10.1007/978-3-319-11370-8_21
41. Gergely, T., Ury, L.: Mathematical foundation of cognitive computing based artificial intelligence. In: Osipov, G.S., Panov, A.I., Yakovlev, K.S. (eds.) Artificial Intelligence. LNCS (LNAI), vol. 11866, pp. 29–64. Springer, Cham (2019). https://doi.org/10.1007/978-3-030-33274-7_3
42. Valiant, L.G.: Cognitive computation. In: Proceedings of IEEE 36th Annual Foundations of Computer Science, pp. 2–3. IEEE Computer Society Press. Milwaukee (1995). https://doi.org/10.1109/SFCS.1995.492456
43. Cognitive Catalysts. IBM Institute for Business Value (2017)
44. van den Hoven, J.: Data architecture: principles for data. Inf. Syst. Manage. **20**, 93–96 (2003). https://doi.org/10.1201/1078/43205.20.3.20030601/43078.11
45. El-Sheikh, E., Zimmermann, A., Jain, L.C. (eds.): Emerging Trends in the Evolution of Service-Oriented and Enterprise Architectures. ISRL, vol. 111. Springer, Cham (2016). https://doi.org/10.1007/978-3-319-40564-3
46. Donaldson, S.E., Siegel, S.G., Williams, C.K., Aslam, A.: Enterprise cybersecurity architecture. In: Enterprise Cybersecurity Study Guide, pp. 95–133. Apress, Berkeley (2018). https://doi.org/10.1007/978-1-4842-3258-3_3
47. Bradley, R.V., Byrd, T.A.: Information technology architecture as a competitive advantage-yielding resource: a theoretical perspective. IJNVO **4**, 1 (2007). https://doi.org/10.1504/IJNVO.2007.012079
48. Goodyear, M., et al.: Enterprise System Architectures: Building Client/Server and Web-based Systems. CRC Press (2017). https://doi.org/10.1201/9780203757239
49. Martin, P.: Kubernetes resources. In: Kubernetes, pp. 19–22. Apress, Berkeley (2021). https://doi.org/10.1007/978-1-4842-6494-2_4
50. Vohra, D.: Docker Management Design Patterns. Apress, Berkeley (2017). https://doi.org/10.1007/978-1-4842-2973-6
51. Masuda, Y., Shirasaka, S., Yamamoto, S., Hardjono, T.: An adaptive enterprise architecture framework and implementation: towards global enterprises in the era of cloud/mobile IT/digital IT. Int. J. Enterp. Inf. Syst. (IJEIS) **13**, 1–22 (2017)
52. Masuda, Y., Shirasaka, S., Yamamoto, S., Hardjono, T.: Architecture board practices in adaptive enterprise architecture with digital platform: a case of global healthcare enterprise. Int. J. Enterp. Inf. Syst. (IJEIS) **14**, 1–20 (2018)
53. Masuda, Y., Viswanathan, M.: Enterprise Architecture for Global Companies in a Digital IT Era: Adaptive Integrated Digital Architecture Framework (AIDAF). Springer, Singapore (2019). https://doi.org/10.1007/978-981-13-1083-6
54. Wang, Y., Towara, T., Anderl, R.: Topological approach for mapping technologies in reference architectural model Industrie 4.0 (RAMI 4.0). In: Proceedings of the World Congress on Engineering and Computer Science, pp. 25–27 (2017)
55. Masuda, Y., Shepard, D.S., Yamamoto, S., Toma, T.: Clinical decision-support system with electronic health record: digitization of research in pharma. In: Chen, Y.-W., Zimmermann, A., Howlett, R.J., Jain, L.C. (eds.) Innovation in Medicine and Healthcare Systems, and Multimedia. SIST, vol. 145, pp. 47–57. Springer, Singapore (2019). https://doi.org/10.1007/978-981-13-8566-7_5
56. Ross, J.W., Sebastian, I., Beath, C., Mocker, M., Moloney, K., Fonstad, N.: Designing and executing digital strategies. In: ICIS 2016 Proceedings, vol. 17 (2016)

57. Ross, J.W., Weill, P., Robertson, D.: Enterprise Architecture as Strategy: Creating a Foundation for Business Execution. Harvard Business School Press, Boston (2006)
58. Gadepally, V., et al.: AI Enabling Technologies: A Survey. arXiv:1905.03592 [cs] (2019)

Organizations Crisis Framework: Classification Crisis Factors with Using Conceptual Modelling

Veronika Vasickova and Vaclav Repa[(✉)]

Prague University of Economic and Business, Prague, Czech Republic
{vasickova,repa}@vse.cz

Abstract. The issue of crisis is a constantly debated phenomenon in many scientific disciplines. The term crisis is considered as a multidisciplinary concept because conceptualization and analysis of this discussed phenomenon remains important for research. The multidisciplinary, manner of managing crisis in different fields and the complexity of examining this problem often lead to the fragmentation of this discipline. A lot of research papers focus on observation and understanding of crisis phenomena, which are aimed at compiling the concept of crisis theory. We react to the fact that an organizational crisis is specified by several attributes with causalities characterized by several crisis life cycle. These life cycles are conditioned by a number of factors that fundamentally affect crisis development. This research paper seeks to understand and explains the concept of organizational crisis through using conceptual modelling. The research goal is to identify crisis factors affecting the development of the crisis and to focus on an idea of a possible identification the crisis tipping point.

To understanding the complexities of the organizational crisis concept, the Methodology for Modeling and Analysis of Business Process (MMABP) is selected for this research. This technique works with a System ontology model represented by the conceptual model with life cycles models of organizational crisis. For conceptual modelling we use standard modelling language UML which contains diagram for sufficient modelling of the Real World modality as well as its related causality (Class Diagram and State Chart).

Keywords: Organizational crisis · Crisis factors · MMAPB · Conceptual modeling · Crisis life cycles

1 Introduction

All efforts to understand the organization crisis are a very scattered field of research where the under-theorization of the crisis concept has led to multiple crisis definition and typology [4, 16, 21, 36]. The monitoring and analyzing of crisis concept are very demanding, especially in terms of multidisciplinary interviewing of the concept, which lead to complications in determining the context [4, 34, 49]. Many research areas and many authors try to capture the crisis concept in the individual frameworks or concepts that identify important factors of this studied phenomenon [1, 4, 15, 34, 49, 50, 56]. A crisis uniqueness is a significant phenomenon that has an impact on the conceptualization

© Springer Nature Switzerland AG 2021
R. A. Buchmann et al. (Eds.): BIR 2021, LNBIP 430, pp. 165–179, 2021.
https://doi.org/10.1007/978-3-030-87205-2_11

of this term [12]. Crisis never occurs in the same way as previous ones. This fact makes its observation difficult and all efforts to validate crisis management procedures cannot be considered as a general approach for describing the concept of a crisis [4].

From historical point of view, the crisis theory has formed based under the mono-lithic methodological approaches dealing with crisis cases: Bhopal [49], Anthrax [6], Chernobyl [11], SARS [29], World trade Center [46] or empirical studies more focused on organizational crises [1, 33, 35, 42, 52, 58]. The importance of these studies lies in the design of a framework for crisis management but also in the general view of an attitude to the crisis. Usually, a crisis is characterized as a negative phenomenon [13, 18, 45, 58]. Negative aspects can be seen mainly in the threat of catastrophe scenarios for society [4]; from the organizational crisis point of view the examples are the financial health of the organization, the loss of its position in the market, the loss of competitive advantage and position or bankruptcy [41, 58] etc. The crisis disrupts normal organizational functions characterized by routine procedures that require standard interventions and interrupt the long-term performance of the organization [40]. Conversely, the crisis situation gives an information about the changes in organization performance and also about the develop-ment of the organization as such (e.g. view on organizations maturity or growth) [1, 55]. The crisis is a part of every organization's development which contains an opportunity to improve and learn from previous crisis experiences [1, 23, 57]. Based on this crisis view, all organizations efforts should be directed at the identification of new challenges in organizations operations and processes [23, 32]. Adams, Phelps and Bessamt [1] noted that crisis brings a turning point which can leads to identification opportunities and gives the opportunity to recovery a malfunctioning system. With this assumption agree Ulmer et al. [56] because crisis can give the opportunity for the organization to be stronger and can be seen as a positive aspect in the field of learning and improving the crisis management techniques [5, 6, 10].

Crisis in its broadest sense, brings a certain summary of historical practices and developments of the system, which at some point will proves to be obsolete. The orga-nization is affected by a crisis if it cannot respond adequately to the situation [23, 26, 38, 58]. With this fact relates the issue of early identification of crisis symptoms lead to the crisis prevention [23]. This issue is in most cases a key factor for successfully resolving or averting a potential crisis [24] and it is discussed by a number of authors who emphasize the proactive style of crisis management [35, 39, 42, 52, 55].

A lot of crisis research and case studies have attempted to create the "manual" for investigating crises. They propose definitions, typologies, models and frameworks for studying organizational crises [23, 31, 35, 53, 58]. Each crisis has an "anatomical" structure and goes through a certain development, which can be characterized by typical features. We are not suggesting that crisis is a necessary precondition for opportunity. However, we highlighted the possibility of intentionality in crisis management and ability manage the crisis factors. The understanding how to manage the crisis, requires the precise knowledge of crisis factors and causalities between them. We would like to extend the background of crisis concept analysis in the field of crisis theory. Therefore, we defined a research question and following research approach: Is it possible to describe the crisis attributes and its life cycles, which are influenced by the crisis factors, using conceptual modeling techniques? First, we studied the crisis organizational framework and introduced them in theoretical background of organizational crisis phenomenon. It is important to understand the problem domain, in this case, the organizational crisis where its solutions are achieved with building of the design artifacts [22]. We agree with arguments of Hevner et al. (2004) that the answer to the fundamental questions for design research is necessary for the basis of the research. The first guideline deals with the utility of the new artifacts where we argued that with using of UML models (class diagram and state diagram) can precisely describe the domain of an organizational crisis. These artifacts brings a new view on organizational crisis as an object with its life cycle that are basic for a consequential modeling of processes. The second guideline focus on an evidence of utility. Proposed artifact in this paper (Organizational Crisis Class Diagram, class diagram and Organizational Crisis Life Cycle, state chart) contributes to mapping of the real world (organizational crisis). MMABP also provides the modeler with the set of rules that help to achieve the needed consistency and completeness of the model, which is another significant contribution of this approach. Second, we characterized selected method for conceptual modelling and studying crisis concept. Finally, we proposed a conceptual model developer under selected method and techniques. Through the selected frame of 4Cs [51] we characterize a key organizational crisis attributes based on literature background and we integrate them into our organization crisis model. This frame was developed by Shrivastava [52] is useful for an identification of crisis factors influencing crisis attributes such as crisis causes, consequences, caution and coping. The frame 4CSs was used in previous research, for instance, has encompassed to important research for forming crisis concept and respecting multidisciplinary of the term crisis [43]. The goal of this paper is an identification possible crisis factor and understanding the role of crisis tipping point. We propose a System ontology model with using a MMABP technique for design of the conceptual model with models of life cycles of selected objects. This type of model respects the important features of the crisis concept, its modality and causality. For the ontology models MMABP is used standard modelling language UML describing the Real World modality as well as its related causality. We introduce the Organizational Crisis Class Diagram and Organizational Crisis Life Cycle state chart.

2 Theoretical Background: Key Perspective of Organization Crisis for Modeling Its Life Cycle

The concept of crisis has its origin in the ancient Greek word "krino", which has the meaning "to assess, measure between two opposite variants, choose or decide". The term "crisis" was derived from this meaning, which was characteristic of the decisive period. The other interpretation has origin in Chinese symbol where one symbols indicate danger, threat and the other sign means opportunity, chance and challenge [14]. The other concept of crisis was recalled in "The Meaning of Crisis" in the field of medical research. The crisis condition is understood as the condition of the patient where there is a turnaround in his treatment. It means that is a period of turbulence leading to transformation among a state [51]. From this metaphor it can be deduced that the crisis is also understood as a decisive and turning point in the development of the subject, when deciding on its future – survival or extinction. This metaphor has borrowed by social scientists for describing crisis in social, political or economic systems. The idea of this paper is based on these historical assumptions, where in each crisis there is a certain break point, which is decisive for its further crisis development. At this point, it is possible to recognize the opportunity that the crisis can often originate. Due to the frequency of possible capturing of crisis concept, the 4Cs framework develop by Shrivastava [50] was selected for the purpose of recognizing the crisis attributes. 4Cs frame suggests that organizational crisis can be characterized by four crises key aspects: causes, consequences, caution and coping. *Causes* include the failures or problems that triggered the crisis and conditions leading to failure [51]. Some authors interpret the triggering event as critical problems leading to inevitable organizations change [1]. A lot of crises are caused by ignoration of warning signals [3, 8] which create the conditions for a crisis origin. Milburn et al. [32] agrees with this fact and noted that crisis occurs when the organization is not able to adequately respond to anomalies caused by the organization´s environment, both internally and externally. *Consequences* are immediate and long-term impacts [51]. However, as stated above, the crisis can be seen as the opportunity and it depends on time horizons and rate of crisis development. The late identification of crisis and the late implementation of crisis measures indicate the crisis escalations with negative consequences [58]. *Caution* is interpreted as a prevention where the taken measures focus on mitigate or minimize impact of the potential crisis [51]. These aspects have a features of proactive crisis management were research studies and scholars debated about the need to anticipate the crisis management and already analyse it in the period of organization stability [35]. The proactive style has described by many researchers in the broader sense of management with common idea that all activities tend to system analysis of warning signals enabling timely detection of potential crisis and these activities shape a system for timely capturing the crisis development [23, 32, 42, 48]. *Coping* includes the features of reactive management where organizations focuses on identifying the crisis and eliminating it. Reactive crisis management includes activities focusing on identifying the crisis and its elimination [35, 51]. Reactive approach focuses on solving and coping the crisis with the necessity to analyse the organizations situation after recovering from crisis [8]. Below, we summarize general crisis attributes derived from these assumption and other crisis definition that was studied and used them to propose a more complete describing the crisis concepts by modelling (Table 1).

Crandall, Parnell and Spillan [8] argue that every crisis goes through a certain phase, and if this phase is sufficiently analyzed, it is possible to protect of organization from the more fatal crisis consequence. Pearson and Mitroff [40] state that it is possible to detect various form of warnings signals with potential crisis symptoms. At this stage, an analysis of the internal and external environment is essential for clearly identification of potential warning signals [8]. The imbalance in not yet apparent, the organization operations are in normal mode, therefore is important deal with potential more thoroughly [30]. Managers have basically two options, either to capture signals or to ignore them. However, the aim should be to capturing of warning signals and take measures to prevent before potential crisis [43]. Commonly, the authors describe potential stage as the pre-crisis stage where is it possible to identify evident symptoms [8, 15]. Fink [14] describes the first stage of crisis as a prodromal phase, where the it is possible to identify both warning signals of potential crisis, but also to recognize signals that are not completely obvious, but are already visible in the organization. The author points out an importance to capturing

Table 1. Derivation of crisis attributes from crisis definitons with using frame 4Cs

---

Generalization of crisis interpretation

4Cs: Causes

Crisis attributes: triggering event, trigger of change, warning signals, tipping point

References: [1, 2, 4, 8, 9–11, 17, 27, 30, 40, 41, 45, 47, 51, 55, 59]

General crisis causes are psychological causes, social causes; natural and ecological causes; technological causes; financial and economic causes. Causes in terms of growth and maturity models. Causes from the point of view of timely identification of the crisis stage. Crisis as a trigger of change. Successful resolution of the crisis leads to the transition to the next stage of maturity or growth model. Each crisis reaches a turning point that is decisive for its further development. Crisis is not just a manifestation of an unexpected event, but often results in the accumulation of organizational problems that are obvious in processes but often overlooked. Crisis does not occur without warning

---

4Cs: Consequences

Crisis attributes: opportunity, danger, failure, critical problem

References: [1, 2, 4, 7, 14, 18–20, 24, 27, 33, 34, 36, 42, 43, 52, 58]

Crisis is a catalyst for positive organizational change. Crisis brings an opportunity to improve and learn. Crisis is a positive impulse for better organization development. Crisis is a change leads to success. The crisis brings a turning point which will bring new opportunities and open the organization to new ideas. Crisis gives an opportunity to heal a malfunctioning system. Crisis is an opportunity for new goals and also an impulse for change in organization. Crisis threats the financial health of the organization, the loss of its position in the market, or the loss of competitive advantage and position. Crisis disrupts the normal function of the organization. Crisis is probable events disrupting future operations affecting the psychological level of the individual or group. The crisis captures an unstable time that brings decisive changes. The organization solve a crisis in the area of its "most pressing" problems

---

4Cs: Caution and Coping

Crisis attributes: crisis management, early identification, prevention, recovery

References: [1, 3, 19–21, 25, 30, 33, 35, 39, 40, 42, 51, 57]

Each crisis goes through a certain phase and if this phase is sufficiently analyzed, it is possible to protect organizations from the more fatal consequences of the crisis. When the organization of the crisis intervenes, the main goal is to minimize the effects of the crisis on stakeholders. During the crisis situation decisions must be made swiftly. Early identification of crisis symptoms is almost a key factor in successfully resolving or averting a potential crisis. Crisis management is a process of capturing and evaluating the warning signals of a potential crisis. Crisis management is a continuous process beginning with prevention by the organization and ending with organisational learning. Crisis management is a set of procedures and principles stabilizing the organization activities. Proactive approach highlights the necessity of anticipating and analysing a crisis during an organisation´s stability. Crisis must be solved immediately

the symptoms and analyzing the possible causes of the crisis. We have introduced the concept of latent crisis, where the organization's effort should be aimed at revealing already apparent symptoms and early detection of them. If the organization wants to averts a crisis and solve it successfully, it is necessary to implement the measures in time [31, 41]. Ignoring the potential and latent phase tends to acute phase [14] where he symptoms are obvious and crisis significantly affects the organizations operations [43]. Some authors argue that the crisis enters to point where is necessary to activate crisis measures and strive for minimalization of damage [42, 59]. This phase requires fast and immediate activations of preventive measures and reductions of crisis consequences [31, 41]. From the maturity and growth models' point of view [17, 47], where the crisis is a trigger of change, is key to identify a tipping point [1]. Every organization goes through certain problems that escalate in a crisis. If the organization recognizes the crisis tipping point in time, it is possible to capture an opportunity in crisis or in the other hand, the crisis consequences probably lead to escalation of crisis or organizations failure [1, 47]. The escalating crisis leads to the final stage of crisis development. The chronic stage is characterized with organization effort for the temporary recovering of basic processes and restoring, returning the organization to balance [30]. According to Fink [14], crisis management plays an important role in this phase, which depends on solving the previous crisis stages and evaluating the procedures. In the event of a successful crisis management, the organization should seek to learn from the crisis experience, to review current plan prevention measures, and to involve stakeholders in the learning process [2, 8, 42, 51]. To clearly description of these characteristic of the crisis, four crisis types resulting from the organizations activities implemented according to the individual stages, were generalized. We used for modeling the potential, latent, acute and chronic crisis which is based on the individual crisis frameworks and model of the mentioned authors. The key role has a crisis learning leading to gaining experience from crisis management process.

3 Method for Studying Organization Crisis Concept

For studying the organization crisis concept, we use the instruments of MMABP methodology (Methodology for Modelling and Analysing of Business Process) [45]. MMABP is a general methodology for modeling business systems. In this conception, business system consists of mutually collaborating business processes that are altogether focused on achieving particular business goals. Achieving of business goals is generally determined by the general rules of the environment in which the processes operate – so called business rules. In MMABP, we model the business rules via modeling the essential concepts of the business system, their essential relationships, and also the essential causality of the system. Except of the causality, this approach is covered by the traditional conceptual modeling. To model the causality, we complete the traditional conceptual modelling with modeling so-called "life cycles" of selected essential objects that the concepts in the model identify, using the UML StateChart (see below). Modeling the causality allows us to directly connect the model of business environment with the model of business processes. In this way, we are able to model both the causality of the business system as well as intentionality of the behaviour of business actors. MMABP uses two UML diagrams that are suitable for modeling the contents of the real-world system in terms of modal logic. It is based on the theory of conceptual modeling and ontology engineering, which we extend with the model of life cycles of selected objects using the UML State Charts. Life cycles allow us to model also the relevant pieces of the real-world causality by a dynamics-oriented model. Real-world causality is exactly what we need to capture for a clear and complete understanding of a general evolution of an organizational crisis.In particular, MMABP works with two basic kinds of models of the business system:

- System ontology model represented by the conceptual model with models of life cycles of selected objects.
- Model of system actors' behaviour represented by the business process models.

In terms of MMABP, studying the concept of organization crisis means modeling this concept together with other related concepts with special care of their causality. In other words, for the purpose of the topic of this paper the relevant part of MMABP is the *System ontology model*. System ontology represents the set of models expressing the important features of the business environment - modality and causality. These models are in informatics also called structural models as they are focused on the structure of the business system (which objects it consists of and how they can interact). For the ontology models MMABP uses standard modeling language UML [56] which contains diagrams for sufficient modeling of the Real World modality as well as its related causality:

- *Class Diagram* is used for the static model of the basic modality of a business system in terms of the conceptual model. The conceptual model represents the so-called system model as it describes the whole system of mutually related business objects. This kind of description principally does not allow capturing the temporal structural aspects of the system (i.e., its causality) as it is focused on the common aspects of the whole system while temporal aspects are principally located just to particular elements of the system.
- *State Chart* is used for the model of the causality connected with the business system object in terms of the so-called object life cycle. This kind of description completes the system model with temporal aspects (i.e., causality). Each model is focused on a single object and describes the causality relevant for the given object in the form of its life cycle.

UML defines some basic relationships between both diagrams, which are also a part of the MMABP rules for modeling the business system ontology. MMABP completes these rules with special rules focused on the meaning of the life cycle model as a model of an essential causality of the business system. These rules are used in the models presented below and also influence the discussion in the following section.

4   Conceptual Model of Organizational Crisis

The crisis is characterized by attributes that fundamentally affect organizational operations. In general, it is an imbalance that should leads organizations to take some crisis approaches to avoid a crisis. From the point of view of the crisis consequences, it is possible to see it either as an opportunity for the change in the manner of adopting fundamental changes or as a danger of escalation of the crisis leading to organizational failure in case of ignoring the crisis phenomenon. The understanding of life cycle of individual crisis types derived form crisis phases essential. These crisis life cycle allow a better understating of causal relationships between individual objects. A proposed class diagram of organizational crisis and state diagram of crisis life cycles were designed to identify individual objects related to the crisis attributes and the subsequent description causalities in crisis life cycle. Due to extent of the state diagram on individual crisis types, only the life cycle of organizational crisis and acute crisis is interpreted in this article, as the area of tipping point identification.

The following "Organizational Crisis Class Diagram" characterizes common aspects of the organizational crisis with related object and represents the important features the modality crisis environment. The diagram shows the classes that describe the relationships between class objects. From the point of view of generalization, the organizational crisis is characterized as possible crisis, originated crisis, sudden crisis and probable crisis. As mentioned above, the crisis development goes through several phases, which are crucial for its possible crisis management and future organizations development as such. Within the diagram, were distinguish crisis types the potential, latent, acute, chronic crisis. The phase of "crisis learning" (CrisisLearning) is not neglected, which, as shown in the modeling of the crisis life cycle, is important for averting the crisis and identifying the "crisis tipping point" (CrisisTippingPoint). The dependence of classes and the description of the given dependence with the identification of the multiplicity of individual relations are expressed between the individual diagrams. The organizational crisis has the role of imbalances in this diagram, which is decisive to determining the crisis approach to the crisis; the role of critical issues that significantly affect the further organizations development and its maturity; the role of the opportunity that the organization can recognize; the role of trigger change required of the further organizations development; the role of the organizational crisis, which is affected by the crisis phenomenon, especially its symptoms and the role of the crisis factor, where it is essential to identify the type of crisis and gain crisis experience in the phase of crisis learning. An important class is the "crisis phenomenon" (CrisisPhenomenon), which is based on the generalization of the crisis causes and its consequences. The relationship between a crisis phenomenon and an organizational crisis is specified by "crisis symptom" (CrisisSymptom) characterized by "warning (WarningSymptom) and obvious symptoms" (ObviousSymptom). As mentioned above, early identification of symptoms is an essential part of crisis management for successful coping with a crisis. The class "crisis symptom" (CrisisSymptom) has the role of a crisis factor, which is vital for the identification of the "crisis tipping point" (CrisisTippingPoint), or the decisive period when the crisis tipping point can be captured. Another important class and also and indispensable part of organizational crisis management is the "crisis approach" (CrisisManagement). From the point of view of previous research [19, 23, 42, 58], proactive and reactive crisis approaches are generally recognized, which have the specific aspects for crisis management. These aspects are considered in the operations of the class. The class "crisis tipping point" (CrisisTippingPoint) is considered in the crisis concept form the point of view of its modeling and description in a completely new way, in the sense of identifying the crisis factors, which are essential for crisis tipping point identification. From the diagram it is obvious that the factors for identifying the crisis tipping point are the "crisis approach" (CrisisApproach), the "crisis symptom" (CrisisSymptom) and the "organizational crisis" (OrganizationalCrisis), where it is vital to identify the crisis types consider the phase of learning form the crisis. The last class is "organization" (Organization) which has role of a participant in the organizational crisis. From the point of view of class operations, a change is important for this class, which is influenced by the overall crisis management, especially by the identification of crisis tipping point and the crisis approach. Gaining experience in managing the crisis process is important for organization too (Fig. 1).

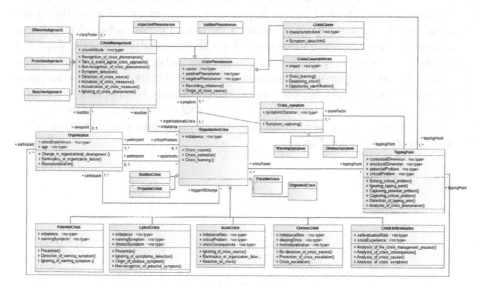

Fig. 1. Organizational crisis class diagram

The following step for proposing a conceptual model of the crisis was the design of an "Organizational Crisis Life Cycle" state chart describing the causality relevant for the given object in the form of its life cycle. Every type of crisis (potential, latent, acute, chronic) has specific aspects conditioned by the previous development of the object states. Thanks to respecting the principles of the UML language [55] and state chart modelling, the life cycles of individual types of crisis were designed. Based on these steps the actions and events to understand the development of the crisis and the possibility to identify the crisis factors essential for analyzation of the "crisis tipping point", were specified. The triggering event in the organizational crisis life cycle is the origin of crisis phenomenon in the form of a certain imbalance in organizational activities. This event is essential for the future crisis development because of taking a stand against the crisis management approach. If the organization deals with the potential symptoms and successfully solve them, then there is no possible crisis escalation and latent crisis. At this stage, the key is learning from the crisis, thanks to the organization evaluates preventive procedures and the manner of identification of potential symptoms. It is clear that the "potential crisis" (PotentialCrisis) is separate life cycle, which is characterized by other specific actions leading either to dealing with the crisis or to crisis escalation to "latent crisis" (LatentCrisis). As the diagram suggests, in the case of ignoring the crisis symptoms or ignoring the crisis phenomenon as such, the crisis escalate through individual type of crisis with the end of organizational failure. In better case, there will be a recovery of organization activities and, thanks to crisis learning and the adopted changes by the organization to crisis mitigation. In each type of crisis, it is characterized by a permanent cycle, which involves a response to the crisis and subsequent crisis learning that improve the crisis management process. An important aspect is the state of a "decisive period" (DecisivePeriod), in which the identification of the "crisis tipping point" (CrisisTippingPoint) which enable to identify a possible opportunity from crisis or, conversely, an escalation of a crisis, plays a key role. In order to achieving this by the organization, it is necessary to clearly identify the critical issues that are detected with crisis management measures. If the organization is able to identify critical problems and capture them, it is possible to avoid an escalating crisis and detect the opportunity in the necessary adopted changes. We assume that it is not possible for the crisis to disappear completely at this stage. At this decisive period, it will only be mitigated in the form of obvious symptoms of a latent crisis. If critical problems are not captured and the necessary response to them is ignored, the crisis escalates into a chronic crisis or the overall organization failure (Fig. 2).

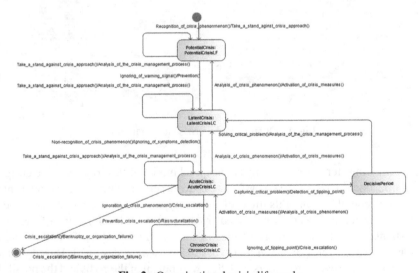

Fig. 2. Organizational crisis life cycle

The life cycle of acute crisis is more specific than others. Therefore, in the following diagram we discuss only "Acute Crisis Life Cycle". Each state chart submodel must respect the number and meaning of the input and output transitions of the represented state of the main state chart model, in this case the Organizational Crisis Life Cycle. Triggering events of "acute crisis life cycle" (AcuteCrisis: AcuteCrisisLC) are actions and event that lead to a crisis approach an analysis of its established procedures, leading to "crisis repression" (CrisisRepression). In this state, the crisis phenomenon is reanalysed and subsequent crisis measures are updated, which take place within permanent cycles in the organizational Crisis Life Cycle model. Event affected by the possible identification of "crisis tipping point" (CrisisTippingPoint) enter this state. In acute crisis, there is a state "origin tipping point" (OriginTippingPoint), where critical problems are captured and crisis measures are activated to mitigate the crisis. This situation also carries the perspective of identifying opportunities based on the analysis of the "crisis phenomenon" (CrisisPhenomenon), where the causes and consequences of the crisis are thoroughly analysed. On the other hand, the analysis of the crisis phenomenon can also identify a deepening crisis that leads to the escalation of an acute crisis, lead in to the organization failure. Ignoring the breaking point of the crisis, the crisis stages its last stage of development, the chronic phase. The submodel captures the fact that even if the organization does not immediately recognize the crisis phenomenon, it is still possible to mitigate the crisis, or analyse the opportunity. This event can occur only with the condition of capturing critical problems in the decisive period and identifying the turning point of the crisis. Otherwise, the crisis escalates again (Fig. 3).

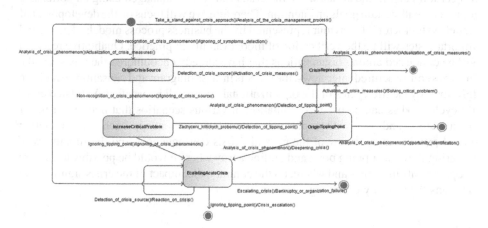

Fig. 3. Acute crisis life cycle

5   Discussion and Conclusions

The proposed model enables us described the main crisis attributes with the modality of objects. Class diagram identify the main crisis factors leading to identification of "crisis tipping point". As with the disease [50], there is a turnaround in times of crisis which will show organization further its developments. In this case, is a possible identification of a "crisis tipping point" leading to an analysis of opportunity, but also to a deepening crisis. Crisis factors such as crisis approaches, organizational crisis (potential, latent, acute and chronic crisis) and crisis symptom should be used appropriately for its identification and timely reaction on critical problems. We assume that a crisis can be avoided in time, especially by catching a potential and latent crisis. The condition of "crisis learning" contributes to this assumption, when the organization constantly evaluates the "crisis phenomenon" and updates its crisis measures. The "crisis tipping point" can only be identified when certain problems arise that require the necessary change in the organization. This idea assumes of maturity [47] and growth models [17], where the organization goes through both the evolutionary phase and the revolutionary phase. It is the revolution that is caused by the crisis, which requires the necessary change [1, 17], both in the organizations processes and often in its overall thinking.

The steps for future research are straightforward. Firstly, these models have to be validated with empirical research. For the validation will be used the quantitative data processing form the questionnaire survey, which will determine the significance of the crisis attribute and verify its proposed life cycles. The results of the research will be compared with a model that will be elaborated and adapted with respect to the results. Secondly, we will focus more on this idea, where an organization can basically reach a certain "crisis maturity", when it will be able to manage the crisis with a certain intentionality based on the experience of the crisis and learning from the crisis process. This intention could be directed, for instance, to an opportunity that will be analysed at a certain point in the crisis, when the crisis can still be managed using an "offensive approach" with certain goals an intention. Therefore, we will focus on the development of model system actors' behaviour represented by the business process models. The second important step will be the verification of model validity in practice. Qualitative research will be conducted among organizations that have already encountered the crisis. On the basis of semi-structured interviews, the objects in class diagram and causalities which are described by the life cycles of the organizational crisis, will be verified. Validated crisis life cycle models can be used to simulation of various scenarios that may occur under given system conditions with respect to the crisis management approach. The simulation should provide an overview of the possible crisis consequences and the timeliness of capturing the crisis tipping point and outline a way where it would be possible to capture an opportunity in a crisis and where, on the contrary, the impact of the crisis significantly threatens the viability of the organization.

References

1. Phelps, R., Adams, R., Bessant, J.: Life cycles of growing organizations: a review with implications for knowledge and learning. Int. J. Manage. Rev. **9** (2007). https://doi.org/10.1111/j.1468-2370.2007.00200.x
2. Alpaslan, M., Green, S.E., Mitroff, I.I.: Corporate governance in the context of crisies: towards a stakeholder theory of crisis management. J. Contingencies Crisis Manage. **17**, 38–49 (2009). https://doi.org/10.1111/j.1468-5973.2009.00555.x
3. Augustine, N.R.: Managing the crisis you tried to prevent. Harvard Bus. Rev. **73**(6), 147–158 (1995)
4. Bertrand, R., Lajtha, C.: A new approach to crisis management. J. Contingencies Crisis Manage. **10**(4), 181–191 (2002). https://doi.org/10.1111/1468-5973.00195
5. Clair, J.A., Dufresne, R.L.: Changing poison into medicine: how companies can experience positive transformation form a crisis. Organ. Dyn. **36**(1), 63–77 (2007). https://doi.org/10.1108/hrmid.2007.04415fad.010

6. Clark, T.: The behavior of common stock of bankrupt firms. J. Finan. **38**, 489–504 (1995). https://doi.org/10.1111/j.1540-6261.1983.tb02257.x
7. Coombs, W.T.: Crisis management and communications. Institut Public Relat. **2007**, 1–17 (2007)
8. Crandall, W.R., Parnell, J.A., Spillan, J.E.: Crisis Management: Leading in the New Strategy Landscape. SAGE Publicaions, Inc., New York (2014)
9. Davidsson, P., Achtenhagen, L., Naldi, L.: Research on small firm growth: a review. Eur. Inst. Small Bus. 1–27 (2005)
10. Deschamps, I., Lalonde, M., Pauchant, T.C., Waaub, J.P.: What crisis could teach us about complexity and systemic management: the case of the Nestucca oil spill. Technol. Forecast. Soc. Chang. **55**(2), 107–129 (1996). https://doi.org/10.1016/S0040-1625(96)00206-5
11. Denis, H.: La reponse aux catastrophes, in Quant I impossible survient. Presses Internationales Politechnique, 507–517 (1993)
12. Dufort-Roux, C.: Is crisis management (only) a management of exceptions? J. Contingencies Crisis Manage. **15**(2), 105–114 (2007). https://doi.org/10.1111/j.1468-5973.2007.00507.x
13. Dwyer, D., Stein, R.: Inferring the default rate in a population by comparing two incomplete default databases. J. Bank. Finance **30**(3), 797–810 (2006). https://doi.org/10.1016/j.jbankfin.2005.11.001
14. Fink, S.: Crisis Management: Planning for the Inevitable. I Universe, Lincoln (1986)
15. Fink, S.: Decision making in crisis: the piper alpha disaster. In: Managing Crisis: Threats, Dilemmas, Opportunities, vol. 6, pp. 103–118 (2002)
16. Gilpin, D.R., Murphy, P.J.: Crisis Management in a Complex World. Oxford university Press, Inc. New York (2008)
17. Greiner, L.E.: Evolution and revolution as organizations grow. Harward Bus. Rev. (1988). https://hbr.org/1998/05/evolution-and-revolution-as-organizations-grow
18. Gupton, G.: Advancing loss given default prediction models: how the quiet have quickened. Economic Notes, Banca Monte dei Paschi di Siena SpA **34**(2), 185–230 (2005)
19. Gundel, S.: Toward a new typology of crisis. J. Contingencies Crisis Manage. **13**(3), 106–115 (2005). https://doi.org/10.1111/j.1468-5973.2005.00465.x
20. Hart, P.: Symbols, rituals and power: the lost dimention in crisis management. J. Contingencies Crisis Manage. **1**(1), 36–50 (1993). https://doi.org/10.1111/j.1468-5973.1993.tb00005.x
21. Herbane, B.: Small business research – time for a crisis-based view. Int. Small Bus. J. **28**(1), 43–64 (2010). https://doi.org/10.1177/0266242609350804
22. Hevner, A., Park, J., March, S.: Design science in information systems research. MIS Q. **28**(1), 75–101 (2004). https://doi.org/10.2307/25148625
23. Jaques, T.: Reshaping crisis management: the challenge for organizational design. Organ. Dev. J. **28**(1), 9–17 (2010)
24. King, G.: Crisis management & team effectiveness: a closer examination. J. Bus. Ethics **41**, 235–249 (2002). https://doi.org/10.1023/A:1021200514323
25. Khodarahmi, E.: Crisis management. Disaster Prev. Manage. Int. J. **18**(5), 523–528 (2009). https://doi.org/10.1108/09653560911003714
26. Kouzmin, A.: Crisis management in crisis? Adm. Theory Praxi **30**(2), 155–183 (2008). http://www.jstor.org/stable/25610919. Accessed 5 June 2021
27. Lagadec, P.: Preventing Chaos in a Crisis. McGraw-Hill, London (1993)
28. Lerbinger, O.: The Crisis Manager: Facing Risk and Responsibility. Erlbaum, Mahwah (1997)
29. Maunder, R., et al.: CMAJ **168**(10), 1245–1251 (2003)
30. Maynard, R.: Handling a crisis effectively. Nation's Bus. **81**(12), 54–65 (1993)
31. Meyers, G.C., Holusha, J.: When It Hits the Fan: Managing the Nine Crises of Business. Houghton Mifflin, Boston (1986)

32. Milburn, T.W., Schuler, R.S., Watman, K.H.: Organizational crisis. Part I. Definition and conceptualization. Hum. Relat. **36**(12), 1161–1180 (1993). https://doi.org/10.1177/001872 678303601206
33. Mitroff, I.I., Alpaslan, M.C.: Preparing for evil. Harvard Bus. Rev. **81**(4), 109–115 (2003)
34. Mitroff, I.I.: Crisis Leadership: Planning for the Unthinkable. Wiley, New York (2004)
35. Mitroff, I.I., Pauchant, C., Shrivastava, P.: The structure of man-made organizational crises conceptual and empirical issues in the development of a general theory of crisis management. Technol. Forecast. Soc. Change **33**(2), 83–107 (1988). https://doi.org/10.1016/0040-1625(88)90075-3
36. Mitroff, I.I., Pearson, C.M.: Crisis Management: A Diagnostic Guide for Improving Your Organization's Crisis-Preparedness. Jossey-Bass, San Francisco (1993)
37. Pauchant, T., Mitroff, I.: Transforming the Crisis-Prone Organization. Jossey-Bass, San Francisco (1992)
38. Pollard, D., Hotho, S.: Crises, scenarios and the strategic management process. Manag. Decis. **44**(6), 721–736 (2006). https://doi.org/10.1108/00251740610673297
39. Preble, J.F.: Integrating the crisis management perspective into the strategic management process. J. Manage. Stud. **34**(5), 769–791 (1997). https://doi.org/10.1111/1467-6486.00071
40. Paraskevas, A.: Crisis management or crisis response system? A complexity science approach to organizational crises. Manage. Decis. **44**(7), 892–907 (2006). Začátek formuláře https://doi.org/10.1108/00251740610680587
41. Pearson, C., Mitroff, I.I.: From crisis prone to crisis prepared: a framework for crisis management. Executive **7**(1), 48–59 (1993)
42. Pearson, C., Clair, J.A.: Reframing crisis management. Acad. Manag. Rev. **23**(1), 59–76 (1998). https://doi.org/10.5465/amr.1998.192960
43. Regester, M.: Crisis Management. Random House Business Books, London (1989)
44. Rosenblatt, Z., Sheaffer, Z.: Effects of crisis – triggered demographic depletion on organizational change: the case of Izraeli Kibbutzim. J. Contingencies Crisis Manage. **10**, 26–38 (2002). https://doi.org/10.1111/1468-5973.00178
45. Rosenthal, M.S.: The end-oflife experiences of 9/11 civilians. Death Dying World Trade Center **67**(4), 329–361 (2013). https://doi.org/10.2190/OM.67.4.a
46. Řepa, V., Svatoš, O.: Adaptive and resilient business architecture for the digital age. In: Zimmermann, A., Schmidt, R., Jain, L.C. (eds.) Architecting the Digital Transformation. ISRL, vol. 188, pp. 199–221. Springer, Cham (2021). https://doi.org/10.1007/978-3-030-49640-1_11
47. Quinn, R., Cameron, K.: Organizational life cycles and shifting criteria of effectiveness: Some preliminary evidence. Manage. Sci. **29**(1), 33–51 (1983). https://doi.org/10.1287/mnsc.29.1.33
48. Sahin, S., Ulubeyli, S., Kazaza, A.: Innovative crisis management in construction: approaches and the process. Procedia Soc. Behav. Sci. **195**, 2298–2305 (2015). https://doi.org/10.1016/j.sbspro.2015.06.181
49. Sheaffer, Z., Richardson, B., Rosenblatt, Z.: Early warning signals: a lesson from the barings crisis. J. Contingencies Crisis Manage. **6**, 1–22 (2002). https://doi.org/10.1111/1468-5973.00064
50. Sheaffer, Z., Manogrin, R.: Executives' orientations as indicators of crisis management policies and practices. J. Manage. Stud. **40**(2), 573–606 (2003). https://doi.org/10.1111/1467-6486.00351
51. Shrivastava, P.: Crisis theory/practice: toward a sustainable future. Ind. Environ. Crisis Q. **7**, 23–42 (1993). https://doi.org/10.1177/108602669300700103
52. Shrivastava, P., Mitroff, I.I., Miglani Miller, D.A.: Understanding industrial crises. Int. Bibliogr. Soc. Sci. (IBSS) **25**(4), 285–304 (1988). https://doi.org/10.1111/j.1467-6486.1988.tb00038.x

53. Shrivastava, P.: Editorial: Industrial crisis management: learning from organizational failures. J. Manage. Stud. **25**(4), 283–284 (1988). https://doi.org/10.1111/j.1467-6486.1988.tb00037.x
54. Shrivastava, P., Mitroff, I.: Effective crisis management. Acad. Manage. Perspect. **1**(4), 283–292 (1987). https://doi.org/10.5465/ame.1987.4275639
55. Schoemaker, P.J.H.: Multiple scenario development: its conceptual and behavioural foundation. Strateg. Manag. J. **14**(3), 193–213 (1993). https://doi.org/10.1002/smj.4250140304
56. UML: Object Management Group: Unified Modeling Language (UML) specification v2.5.1 (2017). https://www.omg.org/spec/UML/
57. Ulmer, R.R., Sellnow, T.L., Seeger, M.W.: Effective Crisis Communication: Moving from Crisis to Opportunity. Sage, Thousand Oaks (2007)
58. Valackiene, A.: Theoretical substation of the model for crisis management in organization. Inzinerine Ekonomika-Eng. Econ. **22**(1), 78–90 (2011)
59. Wagner, C.G.: Proactive crisis management. Futurist **39**(2), 6 (2005)

Compliance and Normative Challenges

Towards Regulation Change Aware Warning System

Marite Kirikova[1,2(✉)] ⬵, Zane Miltina[1], Arnis Stasko[1], Ilze Birzniece[1] ⬵,
Rinalds Viksna[1], Marina Jegermane[1] ⬵, and Daiga Kiopa[2]

[1] Department of Artificial Intelligence and Systems Engineering, Riga Technical University,
Riga, Latvia
{Marite.Kirikova,Arnis.Stasko,Ilze.Birzniece,Rinalds.Viksna,
Marina.Jegermane}@rtu.lv, Zane.Miltina@edu.rtu.lv
[2] Lursoft, Riga, Latvia
daiga@lursoft.lv

Abstract. Compliance has been a research topic for more than two decades. However, in most cases it has concerned static regulations and possibilities to ensure that organizational business processes and financial matters adhere to specific laws and other regulatory requirements. This paper looks at a different perspective. The research question addressed is a possibility to predict how business entities will be impacted by changes in normative acts (regulations). These changes are detected by the proposed warning system that not only monitors changes in normative acts, but also gives an opportunity to analyse organizational data with the purpose of identifying performance of which legal entities could be negatively impacted by the changes in regulations and, thus, which entities are to be warned regarding the estimated consequences of regulatory changes.

Keywords: Warning system · Normative document analysis · Compliance · Document graphs · Performance analytics · RegTech

1 Introduction

This research has been inspired by the EU Directive (EU) 2019/1023 on preventive restructuring frameworks, on discharge of debt and disqualifications, and on measures to increase the efficiency of procedures concerning restructuring, insolvency and discharge of debt [1] (hereinafter – the Directive). The key aims of the Directive are twofold: first, it is aimed at giving the bankrupt entrepreneurs/companies another chance and the second, it aims at simplifying the access to restructuring measures for viable companies that have faced financial difficulties in order to avoid these companies becoming insolvent. The latter aim can, to some extent, be achieved by establishing one or more early warning systems on a country level for entrepreneurs/companies to alert them that there are certain signals in their financial performance and surrounding environment that they could soon be facing insolvency. For the purposes of this paper, the term "Legal entity" is understood as any form of entrepreneurial activity recognised by the state and that is created by individual or group of individuals for the purpose of conducting business.

© Springer Nature Switzerland AG 2021
R. A. Buchmann et al. (Eds.): BIR 2021, LNBIP 430, pp. 183–194, 2021.
https://doi.org/10.1007/978-3-030-87205-2_12

Legal entities depend on normative acts, and they must ensure their compliance with laws and other regulatory requirements of different (international, national, regional) origin and granularity. In many cases this problem is addressed by compliance management methods and tools [2, 3]. In recent years, the so-called Regulatory Technology has emerged in the financial sector [4] that utilizes a power of artificial intelligence in regulation monitoring and other areas.

One of the challenges of global economy is the number of normative acts the legal entities must comply with and, also, the frequent changes in these acts. The possibilities to estimate the impact of new normative acts or changes in normative acts for single legal entity are limited due to time and knowledge constraints. In this paper, we address the possibility to provide a service that combines several evidences, based on analysis of freely available documents, for helping organizations to be alerted about possible threats to their performance caused by changes in regulations. The research goal here is *to construct the high-level design of early warning system, which would help to inform legal entities about possible treats caused by changes in regulations.*

To achieve the goal, the following research activities were performed:

1. Analysis of related works.
2. Analysis of the scope of available normative acts and company performance documents.
3. Constructing the high-level design of an early warning system.
4. Defining scenarios of the warning system's functioning.
5. Implementing parts of the high-level design to tests its applicability for identifying cases for warning.

This paper extends our research on legal entity analysis [5] with the emphasis on normative documents. Its main scientific contribution lies in combination of legal document analytics with the performance analytics. The brief discussion of related work and the scope of available documents is presented in Sect. 2. The proposed high-level design of the warning system RegWarn is presented in Sect. 3. In Sect. 4, some scenarios of its usage are addressed. In Sect. 5, we evaluate applicability of the high-level design by considering first results of implementation of parts of the high-level design. In Sect. 6 we state our current conclusions and point to the further steps of research.

2   Related Work and the Scope of Available Documents

The assumption behind this research is that the changes in normative acts can negatively impact those legal entities for which the compliance to the new laws or rules can cause problems to maintain the same or achieve higher performance indicators than they were able to have before the regulatory changes. Thus, to solve the problem, a regulation change aware warning system (RegWarn) could be helpful, which could indicate the companies (legal entities) which shall receive early warning notifications regarding possible negative impact of changes in normative acts. The research work related to this issue is briefly considered in Sect. 2.1, while regulatory background of the assumption and the review of the scope of the normative documents considered in this paper are discussed in Sect. 2.2.

2.1 Related Work

Legal entities are obliged to comply with normative acts. There has been a considerable research effort put into finding the ways how to do it with the aid of information technology. Several references to such efforts are amalgamated in [2]. However, compliance alone [3] does not help to answer the research question stated in this paper, as we focus here on the potential impact of the changes in regulations on performance indicators of legal entities. One area closely related to this issue is so called RegTech [4], which is popular especially in financial industry. RegTech usually includes such functions as monitoring, reporting and compliance which are performed with the help of software. There are also simpler solutions, mainly focused on change monitoring in normative documents, for instance, normative reference to the aeronautics and space sector [6]. Document change aware retrieval methods are already available and used [7]. Research work in RegTech shows several areas where decisions are to be made if software systems are built to assist the process of legal act (regulation) and data analysis [4]:

- What methods should be used for regulation access, monitoring and analysis (including what vocabularies and thesaurus and ontologies are to be used, what types of alerts implemented, what tagging methods used, how regulatory knowledge base should be constructed, what machine learning methods should be applied, what natural language processing tools are to be used [8]?
- What are the ways human experts can (must) be involved and how they can be supported by software tools [9]?

These questions are relevant also in our research. However, as mentioned above, we shall go beyond of RegTech, as the purpose of research is to predict the possible changes in legal entity performance in case of changes of regulations. So, besides the regulation analysis, performance analysis tools are needed to meet the stated purpose. This means that one more area of related work concerns the usage of open data [10]. While there are many challenges (volume, cleanness, relevance) and advantages of open data, we are concerned here with finding the relevant sources of openly available data and possibilities to compare data provided by different countries by similar legal frameworks countrywide or (in future research) between countries.

2.2 Available Normative Acts and Other Documents

As noted in Introduction section of this article, the key driver for early warning system establishment is borne out of the Directive [1] and the imperatives stated in the directive are now also incorporated in local regulatory requirements of each EU country.

The Directive states the goals of the early warning systems, but it does not specify any requirements towards the implementation of the early warning system, hence it is left for country to decide on what sources to use for the early warning system, how to best achieve the goals of early warning system and ensure its effectiveness and fit to purpose. Article 2 of the Directive [1] stipulates that "Early warning tools may include the following: (a) alert mechanisms when the debtor has not made certain types of payments; (b) advisory services provided by public or private organizations; (c) incentives under national law for third parties with relevant information about the debtor, such as accountants, tax and social security authorities, to flag to the debtor a negative development".

Besides the Directive, this research relies on normative acts openly accessible in normative acts' repositories. Such repositories are maintained by many EU countries. While their structure is similar, still there are slight differences concerning meta-data, volume, and granularity of these repositories.

Given that indications of the early warning system can be arriving as from the input data from the company (debtor) itself as well as from the entities external to the company, then, in the first case, if data is fed by company itself, the early warning system could be built in a form of company self-assessment tool, however, in the second case, if the early warning signs are to be identified by entities external to debtor, the early warning system should be based on data that are published and officially available. This research concerns the second case.

The officially available company data, such as company financial reports take time to be published and usually need to rely on reporting/publishing regularity as stipulated by the regulatory requirements of respective country. The liabilities information from credit registers and tax payment statistics from state revenue service are the data that could be retrieved from the online databases and would be relatively fresh input for the early warning system. Nevertheless, this is still largely relying on historical data of the debtor and, thus, needs to be supplemented with future looking data. This could be achieved through several aspects. First, there could be financial forecasting models embedded in the system to identify early deterioration of the debtor's financial situation and, second, there should be forward looking analysis of country's regulatory act changes to identify any changes in normative acts that could adversely impact company's financial situation and future solvency.The authors of this paper have identified several financial ratios that could be used over period of 4–5 years to analyze the historical trends of an individual company versus the company's industry based on the NACE code of the company main business activity [11]: namely, Gross Profit, Profit Before Tax, Gross Profit Margin, Current Ratio, Debt to Equity, and Return on Assets. Also, Altman Z-score [12, 13] was calculated to obtain data for comparing with RegWarn estimations. These ratios were selected as they have the best ability to summarize key vitals of the company's financial situation as well as they are commonly applicable to all companies (legal entities) and all industries. In order to consider, in the early warning systems, the business specifics and regulatory requirements of various industries, the industry dimension is introduced in the model.

3   Proposed High-Level Design of RegWarn

The proposed RegWarn high-level design, represented in a simplified form using Archi-Mate language elements, is shown in Fig. 1. The high-level design consists of knowledge repository and six functional components. For the modelling purposes, the enterprise architecture modelling language ArchiMate has been chosen, and the simplification is made by focusing only on the components of the system and not to its context (except of data and knowledge sources). The functional components point to the functions of the system and intended software systems are depicted in each functional component. However, the high-level design allows for gradual implementation so that initially any functional component can be realized also manually. The following functional components are included in the high-level design:

1. Document analysis methods and algorithms, which includes two main sub-components (one for reports analysis, another one – for normative act (regulation) analysis).
2. Regularity identification methods and algorithms for identification of data relevant for further analysis.
3. Company performance prediction methods and algorithms.
4. Suggestion generation and distribution methods and algorithms.
5. Normative document monitoring methods and approaches.
6. Prognosis quality analysis.

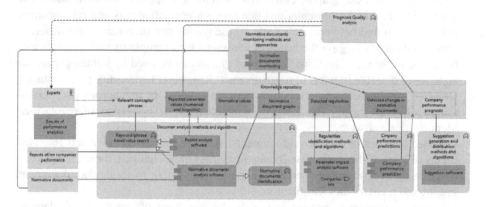

Fig. 1. RegWarn high-level design (simplified).

Functional components 1–4 from the list, are the core of the RegWarn system. Normative document monitoring helps to see the changes in already considered regulations. Prognosis quality analysis is the block that helps to analyse the success of the system in a longer run to see the components, methods, and algorithms that are to be improved or changed.

All above mentioned functions are supported by knowledge repository that consists of both machine-readable data and human readable documents. Different components of the repository can be used by several functional components. The actual relations between the elements of knowledge repository, thus, are not strictly pre-defined, however, different relationships can emerge during the use of knowledge repository components. Some of usage examples are explained in [14].

The following knowledge repository components are used:

1. Relevant concepts and phrases: These are concepts and phrases which can be used in analysis of normative acts (regulations), for instance, as search strings when selecting

target documents. In our experiments we have utilized two origins for obtaining these phrases: (1) expert knowledge and (2) concepts that were included in decision trees of analytical application used for company performance analysis when attempting to classify them in more trustable and less trustable companies [5]. The assumption is that, if the regulatory changes concern these concepts, this normative act's change is the subject for further analysis of its potential impact on performance of legal entities.

2. Reported parameter values (numeric and linguistic). In the previous point, a knowledge component that concerned specific concepts was discussed. Part of these concepts can be analysed with respect to their values also, for instance, profitability as an important concept may be analysed both in regulatory and performance documents with respect to set value of interest.

3. Normative values, these are reported parameter values that have changed in normative acts over the time. The set of these changes is used as an input information for possible normative act change's impact analysis.

4. Normative document graphs. Normative document graphs. Normative document graphs are graphical representations that concern normative acts and relationships between them and between their parts. These graphs can be obtained for different time points depending on the needs of analysis. One example of such a graph will be shown in Sect. 5. The visualizations of graphs can be used by human experts in analysis of possible impacts of normative act changes on legal entity performance.

5. Detected regularities are correlations between changes in regulations and performance of legal entities that have been detected either by expert of regularities identification software.

6. Detected changes in normative documents. When relevant normative acts (the elements of a set of the normative document graphs) have changed, these changes are saved in the repository and serve as alerts for the system to recalculate document graphs and regularities identification.

7. Company performance prognosis is data that has been obtained by calculations and/or from experts. This prognosis points to legal entities which should be advised to pay a specific attention to changes in regulations and predicted problems in their performance.

The RegWarn high-level design assumes availability of external knowledge sources such as domain experts, open and machine-readable repositories of normative acts, and open repositories of legal entity performance reports. The RegWarn high-level design is built by respecting continuous engineering needs in the sense that the system can be digitalized gradually being operational at all stages of development [5]. So, in the beginning it can rely more on the expert knowledge and gradually incorporate new software applications. Some of its usage scenarios are discussed in the next section.

4   RegWarn High-Level Design Usage Scenarios

Based on the selected sources and ratios as described in Sect. 2.2 of this paper, the user interface for company review by the expert (a company card) has been created in the first iteration of RegWarn development (Fig. 2). This card is intended to be populated by data by RegWarn software applications. The card is user interface component that contains all relevant context about a single subject – i.e., company and is comprised of the elements that are necessary for the expert for analysis and decision-making. The company card (Sect. 1 of Fig. 2) contains the key company details to be able to unambiguously identify the company, and also company NACE code for the expert to know to what industry the company belongs to and what industry the company is being compared to. The industry is relevant for two aspects: (1) the reference data that is being used to explain company financial performance and (2) to assess industry specific regulations that company has to be compliant with.

Section 2 of the company card (Fig. 2) is dedicated to the most current data available from external online interfaces and is used to highlight first indications of company financial situation deterioration – missed tax payments to state revenue service/ tax inspectorate of the country and information on delayed loan payments from the credit register.

In order to avoid cases where tax disputes, deviations from calculated and paid amounts or industry specific tax settlements are showing up as false positives, the existence of tax debt is being analysed in context with company turnover – i.e., outstanding tax debt/company turnover. This information is to also be viewed in context of industry comparison as increased tax debt for industry average would be one of the first signals of financial struggles for industry as a whole.

One additional aspect highlighting company's current wellbeing is company's credit worthiness – i.e., company's track record in cover in their loan payments – discipline of paying regular loan instalments based on absence of negative records that banks would be reporting to country's credit register when company has been late with payments as per agreements or had applied for payment grace period.

Section 3 of the company card (Fig. 2) is dedicated to normative acts governing the specific legal entity. This means listing of all recent changes to normative acts that have direct impact on its financial and business performance as well as those normative acts that have been published but yet to enter in the force, as this means foreseeable effect to legal entity performance.

Section 4 of the company card (Fig. 2) depicts six selected ratios of financial performance of a legal entity, which are being analysed over a period of five years to identify the performance trends as well as the Altman Z-score in the historical period over ten years to predict the chances of a business going bankrupt in the next two years [12, 13]. Five and ten years is just one possible sub-scenario of RegWarn calculations. The number of years for calculations can vary, thus, producing several variants of data. Again, the performance of the company under analysis is being compared to the industry average ratios to determine whether the root cause of the performance deterioration is based within the company individual performance or whether this is industry overall case.

These three main blocks of the company card are providing sufficient information to the expert for the company financial situation analysis and highlighting the first signs of financial situation deterioration if such exist.

There are different options for how the system can be used depending on what data/concepts are obtained from expert analysis or by calculation. For instance, if an

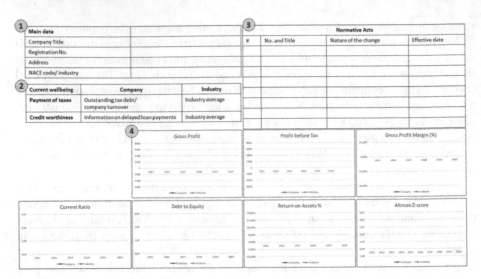

Fig. 2. Company card for the expert.

expert selects relevant normative acts (either directly or by using existing document graphs as described in Sect. 5), the crawlers can check whether these documents have mapping with any of the selected indicators and/or their values; and if yes, then this data is prepared for further expert assessment or algorithms. Each identified relationship between changes in normative documents and the company performance can be saved and later be used for training of machine learning tools. Thus, in different scenarios, the search for regularities (mappings between existing or predicted performance indicator values and changes in normative acts) can be introduced and performed either by expert or by software applications in each functional component, depending on which functions in the system are performed manually and which automatically. Thus, the multimode service (functionality) composition is applied in RegWarn [15].

5   Discussion on RegWarn High-Level Design Applicability

In the current stage of research, we have tested part of the high-level design components presented in Sect. 3, namely, document analysis methods and algorithms, regularities identification methods, company performance prediction methods, and normative documents monitoring methods.

Regarding document analysis methods, we have implemented both – performance report analysis and regulation analysis. We have used openly available data of two countries: the United Kingdom and the Republic of Latvia.

Performance analysis of legal entities was done respecting NACE codes of companies [11]. The details about performed experiments are available in [14]. An example of the result of analysis of the data on raising of sheep and goats is shown in Fig. 3. These results were provided to the expert with the expectation that certain regularities with respect to possible impact of normative documents could be identified, focusing on the year 2019 which shows some consistent changes in the parameters. The changes in normative documents 1 to 5 years ago were seen as potential causes of the changes in parameters. In this case, the expert was not able to identify any regulatory changes which could have caused the changes in performance parameters. At that attempt, the expert was not equipped with the results of normative acts' analysis, i.e., normative document graphs. A normative document graph is illustrated in Fig. 4.

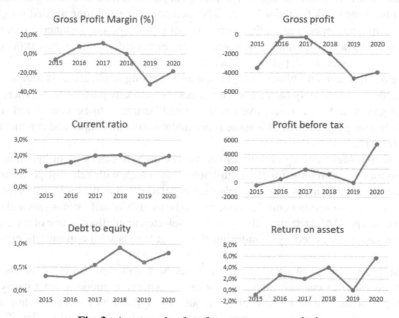

Fig. 3. An example of performance report analysis.

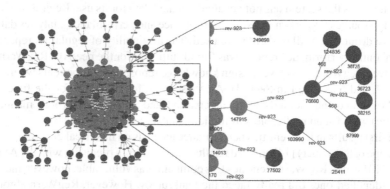

Fig. 4. Regulation (normative document) graph with depth 3 for the Latvian Law "On taxes and duties" at time point 06.04.20 (last update of the document). The numbers of nodes are original identifiers of the regulations.

Figure 4 shows how complex the view on normative documents can be. Light grey nodes show the normative act and its related documents which are in force at the given time point, while dark grey nodes show the related normative acts which are already out-dated. In the next iterations of RegWarn development it is intended to show the normative document graphs to the expert with the possibility to query these graphs with the specific linguistic and numeric items. This could help to identify regulations which might have impacted the legal entities and, thus, build learning data for applying machine learning with the purpose of predicting the possible changes in legal entity performance. Development of normative document graphs is time consuming though not a too sophisticated task for the software. The memory space and time needed for graph development vary depending on the normative act repository. For instance, processing the United Kingdom's normative acts was more resource demanding than processing the normative acts of the Republic of Latvia.

An essential problem, which hinders machine learning application, is that the normative acts are not the only factor that influences the company performance. Therefore, different global and local disruptive events should be traced to be considered in the machine learning algorithms. This issue is not addressed in this paper and is a matter of further research.

The scope of relevant regulations which shall be monitor can be identified manually or automatically, searching for the regulations that match with relevant performance indicators directly or indirectly. For selecting regulations automatically, different thesaurus, ontologies and other concept systems can be used to identify synonyms and other relevant concepts [16]. Detecting changes in the selected regulations is one of the easiest tasks that can be achieved by monitoring these regulations with appropriate crawlers [17].

To sum-up, four out of six (1, 2, 4, and 5, see Sect. 3) functional components have been tested either in manual, interactive or purely automatic mode at least partly. All these components proved to be useful for building hypothesis concerning impact of regulations to the performance of legal entities and for building learning data for the company performance prediction algorithms. The prediction quality analysis component was not tested as the system has not yet amalgamated data for its use. Regarding external knowledge sources (expert knowledge, performance and performance analytics data, and normative documents), all of them were used. The availability of regulation repositories and legal entity performance reports was found sufficient for the needs of RegWarn. Five out of seven RegWarn high-level design knowledge repository components where tested (1, 2, 3, 4, and 6 as listed in Sect. 3). All these were found useful for the functioning of RegWarn, and the methods for their creation and maintenance (both manual and automatic) were found realizable and implementable.

The first results in legal entities performance analysis on historical data (not included in the scope of this paper) [14] showed results which differ in precision with the Altman's Z-score [13] so that RegWarn identifies less companies as vulnerable, however, more than 90% of identified ones did really faced the bankruptcy. However, RegWarn algorithms did not identify all companies that actually had to be warned. Thus, multiple knowledge components and functionality of RegWarn give promising initial results in legal entity performance prediction in the context of changes of regulatory requirements. However, they also show, that the use of expert knowledge will, most probably, play an essential role in the system.

6   Conclusions

In this paper we proposed the high-level design of an early warning system RegWarn, which would help to inform business entities about possible threats to their performance caused by changes in regulations. This design contributes a new approach of using analysis of normative documents and data from financial reports to combine different types of evidence for identifying potential threats to organizations. The high-level design consists of six functional components and knowledge repository with seven components. The first experiments with the components let to draw the following conclusions:

- Good balance between expert and artificially obtained knowledge is needed.
- The impact of other issues but regulation changes must be considered.
- All tested functions of the high-level design proved useful, however more software applications should be evoked to have more convincing results.
- All tested knowledge components of knowledge repository proved to be useful.
- Availability of open data sources in the United Kingdom and the Republic Latvia were sufficient for the experiments.

The experiments exhibited potential for further tests with algorithms that allow using data of one country as learning data for analytics applications with respect to other countries. This possibility has not been explored so far but might be used in RegWarn in later stages of its development. In the nearest future we intend to test the rest of RegWarn high-level design components and then gradually populate the system with new analytical software components depending on availability of (built) data corpuses. In addition, the next further advancements of RegWarn would be evaluation of opportunities for using Blockchain technology that would allow legal entities to make its financial data available in real time. Nevertheless, this is strongly dependent on technology availability and respective country and/or EUsupport and incentivisation or enforcement ability on consistent use of the proposed technology.

Acknowledgments. The research leading to these results has received funding from the project "Competence Centre of Information and Communication Technologies" of EU Structural funds, contract No. 1.2.1.1/18/A/003 signed between IT Competence Centre and Central Finance and Contracting Agency, Research No. 1.19 "Comparative analysis of regulatory and financial data of companies from different countries for forecasting business results".

References

1. Directive (EU) 2019/1023 of the European Parliament and of the Council of 20 June 2019 on preventive restructuring frameworks, on discharge of debt and disqualifications, and on measures to increase the efficiency of procedures concerning restructuring. https://eur-lex.eur opa.eu/legal-content/EN/TXT/?uri=celex%3A32019L1023. Accessed 4 June 2021
2. Gaidukovs, A., Kirikova, M.: Types of linkages between business processes and regulations. In: Rocha, A., Correia, A.M., Costanzo, S., Reis, L.P. (eds.) New Contributions in Information Systems and Technologies. AISC, vol. 353, pp. 343–349. Springer, Cham (2015). https://doi.org/10.1007/978-3-319-16486-1_34

3. Posthuma, R.A.: High compliance work systems: innovative solutions for firm success and control of foreign corruption. Bus. Horiz. (2021). https://doi.org/10.1016/j.bushor.2021.02.038
4. Butler, T., O'Brien, L.: Understanding RegTech for digital regulatory compliance. In: Lynn, T., Mooney, J.G., Rosati, P., Cummins, M. (eds.) Disrupting Finance. PSDBET, pp. 85–102. Springer, Cham (2019). https://doi.org/10.1007/978-3-030-02330-0_6
5. Kirikova, M., Miltina, Z., Stasko, A., Pincuka, M., Jegermane, M., Kiopa, D.: The model for continuous IT solution engineering for supporting legal entity analysis. In: Buchmann, R.A., Polini, A., Johansson, B., Karagiannis, D. (eds.) BIR 2020. LNBIP, vol. 398, pp. 67–81. Springer, Cham (2020). https://doi.org/10.1007/978-3-030-61140-8_5
6. Normative Documentation – Testia. https://www.testia.com/weadvise/normative-documentation/. Accessed 4 June 2021.
7. Sahilu, H., Atnafu, S.: Change-aware legal document retrieval model. In: Proceedings of the International Conference on Management of Emergent Digital EcoSystems (MEDES 2010). pp. 174–181 (2010). https://doi.org/10.1145/1936254.1936284
8. Ferraro, G., et al.: Automatic extraction of legal norms: evaluation of natural language processing tools. In: Sakamoto, M., Okazaki, N., Mineshima, K., Satoh, K. (eds.) JSAI-isAI 2019. LNCS (LNAI), vol. 12331, pp. 64–81. Springer, Cham (2020). https://doi.org/10.1007/978-3-030-58790-1_5
9. Thilakarathne, D.J., Al Haider, N., Bosman, J.: Human-centred automated reasoning for regulatory reporting via knowledge-driven computing. In: Fujita, H., Fournier-Viger, P., Ali, M., Sasaki, J. (eds.) IEA/AIE 2020. LNCS (LNAI), vol. 12144, pp. 393–406. Springer, Cham (2020). https://doi.org/10.1007/978-3-030-55789-8_35
10. Kalampokis, E., Tambouris, E., Karamanou, A., Tarabanis, K.: Open statistics: the rise of a new era for open data? In: Scholl, H.J., et al. (eds.) EGOVIS 2016. LNCS, vol. 9820, pp. 31–43. Springer, Cham (2016). https://doi.org/10.1007/978-3-319-44421-5_3
11. Eurostat: NACE Rev. 2. Statistical classification of economic activites in the European Community. https://ec.europa.eu/eurostat/documents/3859598/5902521/KS-RA-07-015-EN.PDF. Accessed 6 June 2021
12. Altman, E.I.: Corporate Financial Distress: A Complete Guide to Predicting, Avoiding, and Dealing With Bankruptcy, Wiley, New York (1983)
13. Altman, E.I.: Financial ratios, discriminant analysis and the prediction of corporate bankruptcy. J. Finance. **23**, 589–609 (1968). https://doi.org/10.1111/j.1540-6261.1968.tb00843.x
14. Stasko, A., Birzniece, I., Keberts, G.: Development of bankruptcy prediction model for Latvian companies. Complex Syst. Inform. Model. Q. **22**, 45–59 (2021)
15. Rudzajs, P., Kirikova, M.: Variability handling in multi-mode service composition. In: Proceedings of 2nd International Conference on the Human Side of Service Engineering 2014, pp. 1–10 (2014)
16. Rudzajs, P., Kirikova, M.: Towards monitoring correspondence between education demand and offer. In: Linger, H., Fisher, J., Barnden, A., Barry, C., Lang, M., Schneider, C. (eds.) Building Sustainable Information Systems, pp. 467–479. Springer, Boston, MA (2013). https://doi.org/10.1007/978-1-4614-7540-8_36
17. Kumar, M., Bhatia, R., Rattan, D.: A survey of Web crawlers for information retrieval (2017). https://onlinelibrary.wiley.com/doi/full/10.1002/widm.1218, https://doi.org/10.1002/widm.1218. Accessed 6 June 2021

Comparative Study of Normative Document Modeling and Monitoring

Rinalds Vīksna[1]([⊠]) [ID], Marite Kirikova[1] [ID], and Daiga Kiopa[2]

[1] Department of Artificial Intelligence and Systems Engineering, Riga Technical University,
Riga, Latvia
rinalds.viksna@edu.rtu.lv, Marite.Kirikova@rtu.lv
[2] Lursoft, Riga, Latvia
daiga@lursoft.lv

Abstract. In this paper, we model normative document repositories and normative documents of two countries – the Republic of Latvia and the United Kingdom and suggest a framework for tracking normative documents. The framework is used for the retrieval of normative documents, their analysis, and their linking to the normative documents already present in the storage module of the framework. We propose splitting normative documents into parts to allow separate tracking of changes in sections and paragraphs for the purpose of normative document monitoring. Information Retrieval, thus, can be done at an article, a paragraph, or a document level.

Keywords: Legislative acts · Monitoring · Information retrieval

1 Introduction

Monitoring changes in laws and other regulations and their effect on the enterprise is an important topic for every organization. In the financial industry, governance, risk and compliance (GRC) and regulatory technology (Regtech) systems are now standard tools in their compliance departments [1]. Typically, a compliance regimen must include three interrelated but distinct perspectives on compliance, namely, corrective, detective, and preventative [2], where the corrective perspective deals with monitoring of a legal environment and keeping track of the current legislation which is constantly changing. The severity of the problem can be illustrated by the number of legislative acts (regulations) issued per year (see Fig. 1).

Most GRC and Regtech applications are dedicated to the financial sector. It is not possible to directly transfer these applications to other sectors due to a considerable difference between financial legislation and regulations in other areas. Also, the needs and purposes of the financial sector differ from many other areas, especially those related to manufacturing, individual businesses, and the like.

In many countries electronically published regulations are considered legal documents. For instance, in the Republic of Latvia, where normative documents that were published electronically previously used to have only informative value, now, starting

R. A. Buchmann et al. (Eds.): BIR 2021, LNBIP 430, pp. 195–203, 2021.
https://doi.org/10.1007/978-3-030-87205-2_13

from 1st July 2012, are considered official if published on the website https://www.ves tnesis.lv. In this paper, we will consider two electronically maintained sources of legal documents: (1) https://likumi.lv legislative portal, which provides access to legislative acts and the Constitutional Court decisions of the Republic of Latvia, and (2) https://legislation.gov.uk – similar electronic document portal in the United Kingdom. The possibility to use electronic normative documents opens room for the use of new IT applications for monitoring changes in such documents [3].

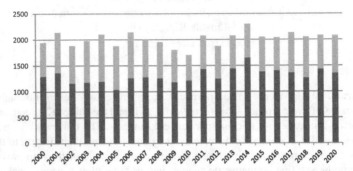

Fig. 1. Count of EU Legislative acts issued in years 2000–2020. Red bars – basic acts, green – amended acts. (Color figure online)

In this paper we seek for answers to the following two research questions:

1) What methods and algorithms can be used for modeling and monitoring of normative documents?
2) What are the essential, country-specific, differences in how the modeling and monitoring of normative documents can or should be performed?

The research focuses on those normative documents, which are not financial institution specific. We use the concept "normative document" to refer to documents that reflect legislative acts or other regulations. The purpose of this paper is to demonstrate the results of regulatory document monitoring and modeling methods application to documents of two countries, so that it would help to distinguish between generic and country-specific issues (if such exist) in normative document monitoring and modeling. The research approach used in this paper is of experimental nature. It is a part of a larger research project on normative document analysis. The results obtained (models, visualizations, and answers to the two stated research questions) are applicable and can be extrapolated to electronically published regulations only.

The paper is organized as follows. The background of the research is discussed in Sect. 2. The models of normative document corpuses of the Republic of Latvia and the United Kingdom are made and compared in Sect. 3. The document monitoring approach is demonstrated and discussed in Sect. 4. References to some related works are threaded in Sects. 3 and 4. Section 5 concludes the paper by summarizing the lessons learned from the experiments.

2   Background

In this section methods and approaches that were, to a larger or smaller extent, used in the research experiments are briefly discussed.

Rudzājs and Bukša in [4] provided a high-level architecture of the Document Analysis and Change Detection system which could be used for the retrieval of regulations and document analysis and preparation for normative document linkage to business processes. In this system, the detection of changes and linking to business processes was realized by splitting regulations into smaller sections and linking business processes to respective sections. If any changes were made to the relevant section, a user could be notified to decide on updates.

Sahilu and Atnafu [5] analyzed Ethiopian legal documents and found that legislations are often complex and are changing over time, forcing organizations to deal with a massive number of normative documents and their updates, which must be integrated into the organization's internal documentary corpus. The model proposed by Sahilu and Atnafu extracts the content of the document and related documents (substitutes, substitutedBY, interpretedBY, and other relations). Documents, which are updated versions of their older versions, can be reranked higher and presented as being more relevant to the user's query.

Lin et al. [6] performed text mining in order to automatically classify normative documents referring to each legislator using the SVM method. The difference between labeling contents by domain experts and the general public was also evaluated to lead the system to real-world application. Accuracy of the classifier was up to 0.73 for target category (24 categories) prediction, however, for topic prediction (4 topics) accuracy was 0.74.

Chalkidis et al. [7] explored the task of regulatory information retrieval (REG-IR) which, given a document describing some procedure in a company, aimed to retrieve all the relevant regulations and vice versa. Chalkidis et al. did a document-to-document information retrieval on a corpus consisting of the United Kingdom's regulatory acts and EU regulatory acts, because the corpus of texts from organizations is difficult to obtain. The task was then to find matching relevant documents from the United Kingdom's corpus given a single EU document as a query. This task was made difficult, by the frequently amended EU directives, which were pre-filtered and post-filtered using the date as a parameter.

Tamilselvam et al. [8] explored the use of news articles (containing regulatory changes) in Compliance Change Tracking by categorizing the articles and, thus, significantly reducing the manual effort spent on reading through the changes. Categorization was performed using two classification criteria: actionability (Irrelevant, InformationOnly, ActionRequired) and applicability (determining relevant business process). Best results in both cases were obtained using a logistic regression classifier.

Change monitoring is usually (except for [7]) tailored to work with documents originating from a single country. In this paper, we explore an approach that can be adapted to monitor changes in multiple countries.

3   Comparison of Normative Documents of Two Countries

There are slight differences in how normative document repositories are structured in the Republic of Latvia and in the United Kingdom. In the United Kingdom, "Primary legislation" is the term used to describe the main laws passed by the legislative bodies of the United Kingdom, e.g., Acts of the UK Parliament, Scottish Parliament, Welsh Parliament, and Northern Ireland Assembly. "Secondary legislation" (also called "subordinate legislation") is delegated legislation made by a person or body under the authority contained in primary legislation. Typically, powers to make secondary legislation may be conferred on ministers, on the Crown, or on public bodies. For instance, the Office of Communications (OFCOM) is given such powers by the Communications Act 2003. Case Law is the set of rulings from court judgments that set precedents for how the law has been interpreted and applied in certain cases. Bye-laws is legislation delegated to bodies such as local authorities, operators of transport systems, or public utilities. The application of Bye-laws is usually limited to a particular local area or the operations of a specific public body. Case Law and Bye-laws are not held on https://legislation.gov.uk. Most types of primary legislation (e.g., Acts, Measures, N.I. Orders in Council) on https://legislation.gov.uk are held in "revised" form, i.e., amendments made by subsequent legislation are incorporated into the text.

In the legislation of the Republic of Latvia, the main types of legislative acts are laws, regulations of the Cabinet of Ministers (Ministru Kabineta noteikumi), and binding regulations of local authorities. The official gazette "Latvijas Vēstnesis", publishes legislative acts and official information.

3.1   Analyzing Electronically Available Normative Documents

We consider a single legislative act (normative document) as a document that has a set of the following attributes: (1) content, (2) metadata, and (3) set of related documents. Their representation is illustrated in Fig. 2.

Fig. 2. Representation of normative document on https://legislation.gov.uk and https://likumi.lv websites.

Content is the textual content of a normative document, indicated with "1" in Fig. 2. On https://likumi.lv website, the dedicated info box containing metadata (id, title, status, validity date of the document, etc.) is available, while, on https://legislation.gov.uk, metadata is scattered in multiple places (see Fig. 2). Https://likumi.lv contains extensive lists of related documents unlike https://legislation.gov.uk website, which contains only a few types of related documents under the "More Resources" tab ("3" in Fig. 2).The general structure of normative documents is represented by the model in Fig. 3 on the left, while attributes of normative documents as found in https://likumi.lv and https://legislation.gov.uk are shown in the middle and on the right, respectively.

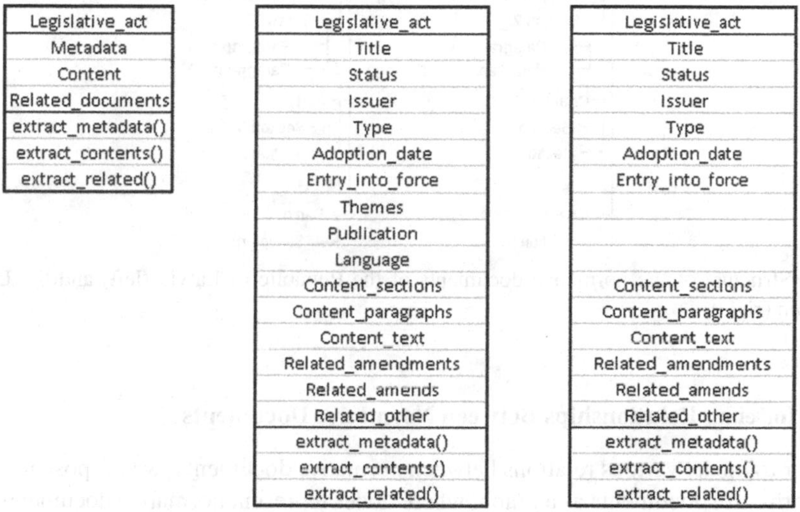

Legislative_act	Legislative_act	Legislative_act
Metadata	Title	Title
Content	Status	Status
Related_documents	Issuer	Issuer
extract_metadata()	Type	Type
extract_contents()	Adoption_date	Adoption_date
extract_related()	Entry_into_force	Entry_into_force
	Themes	
	Publication	
	Language	
	Content_sections	Content_sections
	Content_paragraphs	Content_paragraphs
	Content_text	Content_text
	Related_amendments	Related_amendments
	Related_amends	Related_amends
	Related_other	Related_other
	extract_metadata()	extract_metadata()
	extract_contents()	extract_contents()
	extract_related()	extract_related()

Fig. 3. UML class diagrams for a generic normative document (left), normative document as found on https://likumi.lv website of the Republic of Latvia (middle), and normative document as found on https://legislation.gov.uk website of the United Kingdom (right).

The content section of a normative document is semi-structured; it is structured into chapters, articles, sections, and paragraphs, forming a tree structure. Paragraphs may be updated, removed, or added using separate legislative acts, called amendments, at any time. The amendments to normative documents also have a tree structure, with numbered articles referring back to the document, which is being changed, specifying in which paragraphs and articles the changes are made. The numbering of articles, sections, and paragraphs makes it possible to uniquely identify specific parts of the normative document and keep track of changes made by amendments to specific articles and paragraphs. Document analysis by splitting the documents into constituent parts and structural elements has been used by [4] to gain the ability to refer to a specific article or paragraph which is relevant to a business process.

Legislative acts of both countries are organized according to a tree like structure (Fig. 4), where a document contains multiple numbered sections, which consist of paragraphs (also numbered). In this manner, each section is uniquely identified. Changes to legislative acts are done by separate legislative acts, which specify which sections and paragraphs are changed and what the nature of the changes is (edited wording, bring into force, omit, etc.). By splitting a normative document into constituent parts, it was possible to track each part separately [4], while other approaches to legislation change tracking mostly deal with changes by analyzing textual content on a document level [5, 7, 8].

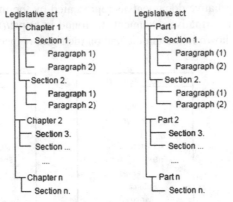

Fig. 4. Structure of the normative documents of the Republic of Latvia (left) and the United Kingdom (right).

3.2 Modeling Relationships Between Normative Documents

In order to model various relations between normative documents, we propose to represent normative documents as a graph, where nodes represent normative documents, and edges are relations – amends, amendedBy, interprets, and others, similarly to [5].

Visualization of normative documents of the Republic of Latvia related to taxes is shown on the left in Fig. 5. As we can see, most of the documents are connected to each other, with only a few documents related to specific events or projects being disconnected from other documents.

On the right in Fig. 5, the "On Taxes and Duties" act is visualized. This is a very well-connected document; therefore, we consider only two types of links: "Amendments" and "Legislative acts, which have changed status" and the relationship depth of 3.

Changes to existing normative documents of both, the Republic of Latvia and the United Kingdom, legislations are made using new legislative acts (amendments), which have the same structure as any other normative document (i.e., they consist of sections and paragraphs) describing changes to be made to other legislative acts. Amendments cannot be made to amendments; changes can only be made to the base legislative acts [9].

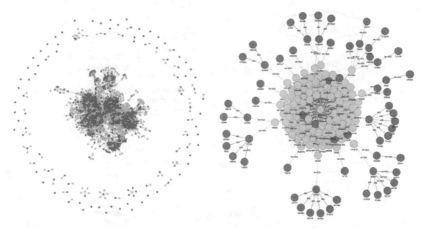

Fig. 5. Left – documents related to taxation from https://likumi.lv. Green nodes are "currently in force", pink – "obsolete"; Right – "On Taxes and Duties" act with related documents. Green nodes are "currently in force", pink – "obsolete". The visualization is made using Python 3.8.5 and pyvis.network package (https://pyvis.readthedocs.io/en/latest/). (Color figure online)

4   Towards the Monitoring of Normative Documents

Respecting frequent and continuous changes in normative documents, an important task is to ensure a possibility to learn about these changes and their potential impact on business processes [4] or other areas of compliance. Therefore, monitoring of normative documents becomes a necessary component of regulatory technology. The monitoring of normative documents is based on their tracking. To enable tracking of legislative environment we propose a framework reflected in Fig. 6. The framework can ensure the normative document tracking that also considers the relationships between normative documents (see Sect. 3). In Fig. 6, the framework is illustrated to be applied for business process compliance, however, it is not restricted just to this field of application and can be used for other tasks where detailed knowledge about regulations is essential, for instance, in the tasks of supplier and partner selection. The proposed framework prescribes several modules that are briefly discussed in the remainder of this section.

The document retrieval module (see 1 in Fig. 6) deals with retrieval of normative documents from various sources – official legal document repositories (such as https://likumi.lv or https://www.legislation.gov.uk), official gazettes (such as https://www.vestnesis.lv/ or https://www.thegazette.co.uk/) or others. For this purpose, a customized crawler [10] written in Python is used. This module also performs monitoring of the data source scanning for new regulations and retrieving new normative documents for the analysis module.

The analysis module performs transformation of raw HTML documents downloaded from the data source into objects described in Sect. 3 by splitting the document into articles and paragraphs, obtaining the list of related documents (if any), detecting the type of the document (a basic act or an amendment), and extracting metadata of the document. Structural analysis is performed using HTML tags which in the Republic of Latvia and in the United Kingdom are used to mark articles (<p id = "p" or < a id =

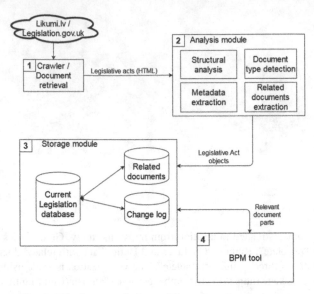

Fig. 6. The framework for tracking normative documents.

"section-5") and detecting paragraphs as subdivisions of articles. Metadata is extracted from relevant document parts using XPath[1] syntax. To obtain a list of related documents, a separate request to the data source is used using the current document id as a search query. Document type detection is done by analyzing text for the presence of specific phrases ("The XXX Act is amended as follows." Or "Izdarīt XXX likumā (…) šādus grozījumus:"). The result of the analysis of a normative document is a Legislative Act object, which is used to update the Current Legislation database.

If the new Legislative Act is a basic act, it is added to the Current Legislation database and, if the new Legislative Act is an amendment, its content is used to update the legislative act it is changing. If an update is made, it is registered in the change log, to allow tracking of the changes. The data is stored internally, relative to the organization, together with other parts of the system to keep the data in-house.

The updated parts are extracted from the text of the amendment semi-automatically. In cases where it is possible to extract the updated article, paragraph, and changed text, it is updated; In cases where only changed article or paragraph is known, but the wording does not allow a simple replacement of a term or a paragraph, the user is notified with proposed changes for review. This module also performs checking if any of the amended parts are referred to by the BPM tool and notifies the user about relevant changes.

5   Conclusions and Future Work

In this paper, we have described an approach for modeling normative documents and a framework for tracking changes of normative documents published on Internet. We

---

[1] https://www.w3.org/TR/xpath-31/.

analyzed normative documents of the Republic of Latvia and the United Kingdom and found that the normative documents of both countries follow the same basic structure and, thus, the same approach with minor modifications can be used for their modeling and monitoring that is based on tracking the changes in the documents.

The framework prescribes that a normative document is split into constituent parts, which are tracked separately, thus, when, e.g., amending law is published, the user is notified only if amended parts are relevant to user-defined rules, otherwise, changes are simply tracked and stored. Extraction of document content (for basic acts) and extraction of amended parts (for amendments) is done using XPath and keywords.

Further work includes the development of the prototype and testing this approach on full-scale legislation datasets from https://likumi.lv and https://legislation.gov.uk. Visualization of the data, user interaction, and reporting is not part of the proposed framework, but could be introduced in the future as a separate module.

Acknowledgements. The research leading to these results has received funding from the project "Competence Centre of Information and Communication Technologies" of EU Structural funds, contract No. 1.2.1.1/18/A/003 signed between IT Competence Centre and Central Finance and Contracting Agency, Research No. 1.19 "Comparative analysis of regulatory and financial data of companies from different countries for forecasting business results".

References

1. Lin, T.: Compliance, technology, and modern finance. Brooklyn J. Corp. Financ. Commer. Law. **11**, 6 (2016)
2. Sadiq, S., Governatori, G.: Managing regulatory compliance in business processes. In: vom Brocke, J., Rosemann, M. (eds.) Handbook on Business Process Management 2. IHIS, pp. 265–288. Springer, Heidelberg (2015). https://doi.org/10.1007/978-3-642-45103-4_11
3. Normative Documentation – Testia. https://www.testia.com/weadvise/normative-documentation/. Accessed 04 June 2021
4. Rudzajs, P., Buksa, I.: Business process and regulations: approach to linkage and change management. In: Grabis, J., Kirikova, M. (eds.) BIR 2011. LNBIP, vol. 90, pp. 96–109. Springer, Heidelberg (2011). https://doi.org/10.1007/978-3-642-24511-4_8
5. Sahilu, H., Atnafu, S.: Change-aware legal document retrieval model. In: Proceedings of the International Conference on Management of Emergent Digital EcoSystems - MEDES 2010, pp. 174–181. ACM Press, New York (2010). https://doi.org/10.1145/1936254.1936284
6. Lin, F.-R., Chou, S.-Y., Liao, D., Hao, D.: Automtic content analysis of legislative documents by text mining techniques. In: 2015 48th Hawaii International Conference on System Sciences, pp. 2199–2208. IEEE (2015). https://doi.org/10.1109/HICSS.2015.263
7. Chalkidis, I., Fergadiotis, M., Manginas, N., Katakalou, E., Malakasiotis, P.: Regulatory Compliance through Doc2Doc information retrieval: a case study in EU/UK legislation where text similarity has limitations (2021). ArXiv preprint: https://arxiv.org/abs/2101.10726v1
8. Tamilselvam, S.G., Gupta, A., Agarwal, A.: Compliance change tracking in business process services (2019). ArXiv preprint: https://arxiv.org/abs/1908.07190
9. Normatīvo aktu projektu sagatavošanas noteikumi. https://likumi.lv/ta/id/187822-normativo-aktu-projektu-sagatavosanas-noteikumi. Accessed 19 June 2021
10. Kumar, M., Bhatia, R., Rattan, D.: A survey of Web crawlers for information retrieval (2017). https://onlinelibrary.wiley.com/doi/full/10.1002/widm.1218

The Impact of the 'Right to Be Forgotten' on Algorithmic Fairness

Julian Sengewald$^{(\boxtimes)}$ (iD) and Richard Lackes (iD)

Technische Universität Dortmund, Otto-Hahnstr. 12, 44227 Dortmund, Germany
{julian.sengewald,richard.lackes}@tu-dortmund.de

Abstract. Enterprises may often deal with situations in which they cannot use a part of their data for machine learning: (1) Privacy laws grant users to determine that their data should not be used any longer by the data holder (i.e., a data record must be removed). (2) Privacy laws may require anonymization, which leads to the removal and masking of specific attributes from a dataset. (3) To avoid ethical issues, it may be necessary to prevent machine learning algorithms from using attributes in their decision-making that may be viewed as discriminatory (i.e., a feature may not be used because it is unethical). Given these challenges, we analyze how removing data from an information system in the cases mentioned above affects its predictive performance and the fairness of its decisions.

Keywords: The right to be forgotten · Privacy · Algorithmic fairness

1 Introduction

Personalized recommendations based on machine learning is a widely used technology in digital services (for example, displaying personalized news or providing users' favorite music based on their preferences). At the same time, many digital services are built on the assumption that their users will provide data to fuel the machine learning engine, resulting in more accurate personalized recommendation capabilities. However, public opinion on the legal regulation of machine learning systems is changing. Numerous laws governing data gathering and machine learning technology have been enacted, affecting systems that are already in use. For example, the EU General Data Protection Regulation (GDPR) in 2016, the UK Data Protection Act 2018, India's Personal Data Protection Bill in 2018, and the California Privacy Rights Act 2020. This places digital service providers in the challenging position of having to engineer a technical solution to meet the legislation's standards for privacy and personal data, but also to ensure that the personalized recommendations are still useful.

The implications of these laws are potentially not just limited to privacy, because machine learning depends on the data that the individual users provide. Some regulations allow self-selection (tracking, privacy consent) and self-redaction of personal data (e.g., personal data deletion request). In general, self-redaction occurs when consumers request that their data be deleted from a digital service and self-selection occurs when the user chooses how much data is collected. Users may choose to share selectively just certain

© Springer Nature Switzerland AG 2021
R. A. Buchmann et al. (Eds.): BIR 2021, LNBIP 430, pp. 204–218, 2021.
https://doi.org/10.1007/978-3-030-87205-2_14

types of information [1, 2] or to remove [3] (e.g., because of growing concerns about data breaches or increased privacy awareness). Selective data sharing and deletion may also occur if consumers experience discrimination based on the data they provide and prefer instead not to share such information.

In practice, groups of users may only disclose a portion of their digitally available information to a service provider. Some examples are: A person may desire to keep their gender hidden to avoid gender discrimination. Similarly, their ethnic origin [4]. In all of the aforementioned cases, the user will either restrict data sharing or ask the digital service to erase their personal data. For data analysts, all of these possibilities culminate in the fact that they only deal with a subset of available data and that certain data have been manipulated or anonymized.

But, what are the consequences of limited data availability on machine learning model-based decisions in digital services under selective data sharing and different technical solutions for implementing the 'right to be forgotten'? Two factors must be considered: the overall quality of the analytic results (e.g., classification performance), on the one hand, and how well the machine learning systems perform for each demographic group, on the other. The second factor is frequently referred to as algorithmic fairness [5, 6]. We hypothesize that minority groups suffer from limited data availability because the remaining group members cannot fully compensate for a loss of training data, and hence unfairness may arise. Our contribution emphasizes the impact on the fairness of the predicted outcomes. Fairness is an increasingly important concern in developing automated systems and algorithmic decision-making, especially in recent years [7–9]. The paper analyzes the empirical evidence from a simulation study on machine learning automated credit lending decisions to address the research question:

RQ: What is the effect of different technical solutions for implementing the 'right to be forgotten' on algorithmic fairness for minority groups and overall predictive capability in machine learning based decision-making?

The conceptual background on legal requirements for data deletion and anonymization is reviewed in Sect. 2.1 and 2.2, and the cybersecurity problem in machine learning models is reviewed in Sect. 2.3. We then proceed to non-discrimination in automated systems and the special category of data according to GDPR in Sect. 2.4. and the measurement of fairness in Sect. 2.5. Related literature is summarized in 1.1. In Sect. 2, we develop our hypotheses and present the technical details of our analysis in Sect. 4. Section 5 contains the empirical analysis. The paper concludes with a discussion of its results in Sect. 6.

2 Conceptual Background

2.1 The Right to Be Forgotten

Two provisions included in the GDPR grant users the right to control their data once transferred to a digital service. First, the well-known 'right to be forgotten' is founded on Article 17's user's withdrawal of consent for lawful data processing and entails the complete removal of their data. Second, Article 18 gives data subjects the right to restrict the processing of their personal data, but does not entail data removal. However, if the data is restricted from processing, it cannot be used for machine learning any longer and hence has the same result as removing it.

However, strictly deleting all records from the database can cause additional technical burdens [10]. Furthermore, as data deletion processes must comply with a wide range of different regulations and laws to be lawful in all jurisdictions, the complete deletion of a record may be unlawful as enterprises may also be obliged to preserve certain information (e.g., the GDPR itself states some exceptions when data must not be deleted). Anonymization is a reasonable choice for such a circumstance because the impacted records will not be deleted entirely from the database. Instead, all the data about the person is masked in a way that the data record cannot be linked back to the actual individual. The enterprise may pursue different degrees of data masking to be compliant with different legal requirements or regulations. Furthermore, if the data is masked to be considered anonymous, then data deletion requests can be rejected [11].

The GDPR requires data controllers to give their subjects the power to select where and how their data is handled. On the other hand, data controllers can reconcile privacy rights with other organizational goals using anonymization techniques.

Similar regulations to those found in the GDPR can also be found in other laws, such as Article §25 and §27 of the Indian Data Protection Bill [12] or Nr. 1798.105 of the California Consumer Privacy Act [13] and §46–48 of the UK Data Protection Act [14].

2.2 Anonymization

Anonymization removes all the personally identifiable information from a database to prevent identification. The identification risk assessment should consider both current and potential technological developments that will increase the risk of identification. Therefore, enterprises may opt for a higher degree of anonymization to also reduce the risk of identification in the future.

Proper anonymization requires at least the removal of any direct identifiers from the data records. Since direct identifiers are often discarded because they usually do not carry any information that can be used for machine learning, therefore deleting these identifiers does not amount to any information loss for machine learning. However, a subset of attributes may require different handling. Quasi-identifiers (QIDs) are attributes where a single QID may not be sufficient to identify a person, but several QIDs together might be used to identify all or at least some records. In such circumstances, the privacy model of k-anonymity can be employed to suppress or generalize some QIDs as needed to maintain privacy protection [15, 16]. Because QIDs are often used in machine learning models, their distortion may harm machine learning performance.

2.3 Cybersecurity

The cybersecurity of public machine learning models is complicated by the fact that they are typically trained with personal data. The so-called model inversion attacks can expose specific properties of training data (e.g., certain features), whereas membership derivation attacks can reveal whether an individual is a member of the training data [17–19]. Furthermore, the security of machine learning services accessible via an API is vulnerable [20]. Because of this, some researchers doubt a machine learning model's anonymity [18].

Owing to these security concerns and the relationship to privacy laws, enterprises are confronted with the risk implied by storing sensitive data in a machine learning model. As a result, some data may be removed from publicly accessible machine learning models to prevent leakage of sensitive personal data.

2.4 Non-discrimination and Special Categories of Data

Algorithmic fairness seeks to prevent people from being mistreated in automated decision-making based on their membership in a particular group. These groups include, for example, those defined by age, gender, and interests [9, 21]. In practice, many deployed machine learning models have been found to perform substantially differently across various demographic groups [9, 22]. Therefore, businesses may want to avoid utilizing particular groups in machine learning models for ethical concerns. Also, the GDPR and other laws agree that certain sensitive data require specific handling (technical or processual). For example, some categories of sensitive data are not allowed to be processed (e.g., Art. 9 No. 10 GDPR). Similar data types are defined in other legislative systems [12, 14].

Because of the aforementioned regulatory restrictions, organizations may remove sensitive data from their machine learning programs to comply with privacy or non-discrimination regulations.

2.5 Fairness Measurement

Technical fairness metrics quantify the extent of discriminatory treatment of a particular group. These metrics measure how much weight machine learning systems put on protected attributes. Protected attributes are characteristics of a person that should not be used as the basis for decisions as determined by law, courts, other authorities, or moral reasons. The protected attributes can signify membership to particular groups, as stated in Sect. 2.4. The domain of the protected attribute is then the set of all possibly existent groups.

For simplification, we assume that there are only two groups: g_1 is the group for which a possible unfair treatment is of concern relative to the treatment that group g_2 receives. The prediction of the machine learning system is denoted by $h(x)$ for any individual/record x. If $h(x) = 1$ the machine learning system will allocate a benefit based on the information contained in the record x. For example, $P(h(x) = 1|x \in g_1)$ is then the probability of how often the machine learning system will allocate a benefit to the members of the group g_1. We define g_1 as the possibly disadvantaged group. Using this setup, the following fairness metrics can be defined:

The Demographic Parity Metric. The demographic parity metric considers an allocation unfair if it disproportionally allocates a benefit to one community:

$$DP = \frac{P(h(x) = 1|x \in g_1)}{P(h(x) = 1|x \in g_2)} \tag{1}$$

The criterion relates to the disparate impact criterion, also known as the 4/5 rule, in the US labor law [6]. Concretely, if $DP < 0.8$ then a disparate impact (DI) occurs for which a company can be held legally accountable if the allocation can not be justified.

It is possible to broaden the measure's scope to include whether or not the benefit recipient is eligible. For example, in supervised machine learning, this knowledge would be carried by the real class label y(x) where $y(x) = 0$ are non-eligible, and $y(x) = 1$ are eligible for any record x.

Equalized Odds (ED). A fairness metric that conditions the actual outcome predicted by an automated system is the equalized odds constraint [5]. Let the true-positive-rate $TPR(g_1) = P(h(x) = 1|x \in g_1, y(x) = 1)$ and the False-positive-rate $FPR(g_1) = P(h(x) = 1|x \in g_1, y(x) = 0)$. Then equalized odds requires:

$$TPR(g_1) = TPR(g_2) \text{ and } FPR(g_1) = FPR(g_2) \tag{2}$$

Equal Opportunity (EO). Only requiring equal predictive outcome for the positive class leads to the fairness criterion equal opportunity [5]:

$$TPR(g_1) = TPR(g_2) \tag{4}$$

Equal opportunity is a less stringent fairness constraint than equalized odds. Only positive recall must be the same across the groups.

Although disparate impact lends itself to a directly quantifiable metric, equal opportunity and equalized odds are defined as a fairness constraint. To quantify deviation from the optimal fairness situation in the latter case, we define unfairness as:

$$Unfairness^+(g_1, g_2) := \left| 1 - \frac{TPR(g_1)}{TPR(g_2)} \right| \tag{5}$$

For the negative recalls, $Unfairness^-(g_1, g_2)$ can be specified accordingly. Thus, one obtains a metric for measuring how equal opportunity and equalized odds fairness constraints are violated. Using the absolute value in Eq. (4) results in the favoring and disfavoring of the group g_1 over g_2 are weighted in the same way.

2.6 Related Literature

In the context of the right to be forgotten, the term 'machine unlearning' emerged first in a setting where one had to develop an efficient technique to erase data from a huge machine learning system because retraining would have been too time-consuming [23]. Further work investigated the viability of design principles for developing efficient dele-tion algorithms and considered an unsupervised learning problem [24]. Using these principles, the authors created two k-means algorithms for efficient data deletion. Other work was directed towards providing guarantees for linear classifiers that ensure that the machine learning model behaves as if it had never been exposed to the data [25].

Aside from record deletion, some work has been done on removing sensitive features from a machine learning model. In such a setting, robust submodular maximization methods were used to efficiently eliminate features from a model after being trained since it is not always known before model training that a feature is sensitive [26].

Precautionary de-personalization (under the restrictions of privacy laws like the GDPR) is another approach to handling the right to be forgotten. There is a wealth of literature dealing with disclosure control techniques for data publishing (e.g., 27–32).

Finally, some work has been done on implementing the right to be forgotten for complex data structures, such as blockchains or streaming/distributed data [26, 33].

Despite having a wide variety of available technologies, little study has been done to examine how data deletion affects other societal objectives. This is unfortunate as privacy laws may be designed to protect human rights and interests. One of the few studies of unintended consequences of enforcing privacy in machine learning is [34]. They find attacks executed before and after data erasure give an adversary insight into the erased record's sensitive information. Thus, data deleted from the machine learning model is eventually disclosed, resulting in an outcome entirely contrary to what was intended.

3 Hypothesis Development

Enterprises may face different situations in which they need to delete a part of their data from machine learning to comply with privacy regulations. We refer to these strategies as personal data protection strategies (PDDR) strategies. By enumeration, we give four strategies a digital service may pursue on a technical level:

1. **Record deletion**. Deleting the complete record, which may be the most straightforward strategy.
2. **Record masking**. Only masking, anonymizing, the record(s) for which a data removal was requested (compliant with previous jurisdiction [11]). Masking can be seen as a soft deletion in that it erases certain details while leaving other details intact.
3. **Anonymizing the complete table**. Performing proactive anonymization of all records so that the corresponding regulation of data removal does not apply.
4. **Deleting a single sensitive attribute or collection of sensitive attributes**. For example, if such sensitive features are typically often requested to be removed, it may be practical not to use them in the machine learning system at all so that the model does not need to be retrained. Furthermore, the omission of such attributes may also be considered ethical or may mitigate privacy leakage if a cyberattack comprises the machine learning model.

The consequences of PDDR strategies on algorithmic fairness of data deletion in the context of a selective-random PDDR mechanism have gone mainly unexplored. By the selective-random PDDR mechanism, we refer to the case when data deletion requests are overrepresented within a particular demographic group.

The literature investigated different reasons why machine learning models may discriminate. For example, the data being used for machine learning is imperfect by itself, and the machine learning model may learn to reproduce already existing discrimination in the machine learning model [7]. In addition, some features may be less reliable or more inaccurately collected for a part of the population or just not a good predictor for some (minority) groups [7, 35]. Also, feedback loops can degrade the fairness of machine learning. A feedback loop occurs when the machine learning model decisions also affect what training data the machine learning model can see in the future, which leads to 'skewed training examples' [7, 36]. Based on these considerations, we developed our first research hypothesis:

- *H1: Record deletion and masking lead to increased unfairness.*
  When data is deleted, the machine learning model has fewer examples to learn from. If a minority group was already more difficult to learn, since a minority group encompasses by definition fewer instances, eliminating data only exacerbates the issue, increasing existing errors. If more errors occur only in one group, the TPR will decrease for this group and increase overall unfairness. Similar arguments can be made for masking, which reduces the amount of information in a specific value for an attribute. However, we expect the loss of information caused by masking a single QID or a collection of QID to be less than the loss of information caused by removing records:
- *H2: Record deletion causes higher unfairness than record masking.*
  We model two separate scenarios, low and high, to differentiate further the impact of different data deletion mechanisms on fairness. In scenario low, only a small number of deletion requests come from a potentially disadvantaged group (e.g., the machine learning system is actually discriminating or wrongly perceived to be discriminating, and users, therefore, request for the deletion of their data). In scenario high, a larger number of data removal requests stem from the potentially disadvantaged group.
  If people, who are eligible or qualified for a particular benefit allocated by a machine learning model (e.g., university admission, loan application, or job offers), are discriminated against or perceive the possibility of their discrimination because of affiliation with a group, they could request data removal. The ability to request data removal reduces the number of cases the machine learning model can see certain characteristics linked to a particular outcome, leading to more unsatisfactory overall performance for that group. We suppose that data deletion induces a special case of 'skewed' training examples [7]. We thus formulate the following hypothesis:
- *H3: The greater the number of users requesting personal data be removed, the worse the impact on fairness and predictive performance.*
  Anonymizing the entire table has an approximate equal impact on all groups and makes it more difficult for the model to learn from them. That is, for example, the group-specific recall would decline for all groups and thus would not affect primarily a single group-specific recall. However, if there is any discontinuity between group-specific sample size and group-specific error, the recall for a marginalized group may decrease further, which is only partially offset by the comparatively more minor increase in group-specific error (i.e., lower recall) for the other group. In that case, the unfairness may increase slightly:
- *H4: Anonymizing the whole table affects unfairness only to a small extent.*
  Deleting a feature from a table may be closest to the notion of fairness that usually involves human decision-making (e.g., many countries have privacy protection

and anti-discriminatory laws prohibiting companies from requesting photos of job seekers).

However, while this technique may help to minimize unfairness in cases where the machine learning model could use a particular attribute to discriminate against a marginalized group (i.e., the inability to reproduce discrimination that is already existing in the world through the distribution of the label attribute [7]), the effect only takes place if that information can be removed entirely from the dataset. In many cases, this may not be feasible because data is usually multivariate, in the sense that attributes are not only associated with the label attribute but also with each other. That is the case with redundant information. Consequently, the machine learning model may deduce the excluded feature from the remaining features in the dataset [19].

- H5: *Removing a feature from the table may reduce unfairness if no redundant information is contained in the database, and, quite seldom, it will increase it.*

4 Simulation Study

The study was designed to test the hypothesis that deleting a subset of data records, values of a set of attributes of a single record, or entire attributes would lead to a loss of fairness (PDDR strategies). For this, we conducted a simulation study.

4.1 Implementation of PDDR Strategies

In particular, record and feature deletion is just directly implemented by erasure from the dataset. Masking is implemented as follows: Categorical attributes may be masked with a dummy value such as "*", and numerical attributes are imputed (as if they were missing data) using a naïve mean imputation procedure. We randomly include some "*" already in the training data to avoid model retraining. Anonymizing the whole table was carried out by the k-anonymity model on the QID (i.e., generalizing the QID as needed to achieve k-anonymity), for which we used the datafly algorithm [15, 37]. Datafly chooses the QID heuristically for generalization.

4.2 Simulated Scenarios

We utilized stratified 70% subsampling to generate the bootstrap replicates. The same bootstrap replicates are used to compare each of the three deletion techniques. We also repeatedly evaluate different users that could have solicited a request for data deletion on the same replicates. The number of bootstrap resamples is 500, and the number of different sets of users that request data deletion is three, yielding a total of 1500 evaluations. In addition, the performance of the machine learning model is evaluated using ten times repeated five-fold cross-validation. The selective-random PDDR mechanism is implemented as follows: The probability for a user group soliciting a PDDR is $P("solicit PDDR"|x \in g_1) = \tau$ and $P("solicit PDDR"|g_2 \in x) = 0$. We simulated two different parameter values for τ describing the percentage of users who request data deletion: a low scenario (5% of group members solicited data deletion) and a high scenario (20% of group members solicited data deletion).

4.3   Technical Details, Data, and Software Used

We rely on the caret package for R to build a machine learning pipeline (i.e., upsampling, cross-validation, scaling, tuning) [38]. We use the `German credit` dataset from the UCI repository as it is a common benchmark dataset for studies in privacy and fairness research. The QID attributes were age, gender, and foreign, for which we developed a suitable generalization hierarchy. Following a setup in previous literature, the fairness-protected attribute was age ≤ 25 years (young vs. old credit applicants) [8].

5   Empirical Aanalysis

5.1   Bradley-Terry Model

We compared the performance of the machine learning model across different PDDR to address data deletion requests. In particular, we are interested in how predictive performance and unfairness compare across the PDDR strategies and which PDDR strategy is preferable according to those metrics. For this comparison, we used the Bradley-Terry model [39]. In the Bradley-Terry model (BTM) β_i denotes the ability of the PDDR strategy i to achieve better performance on a metric γ that an alternative strategy j. The probability that i wins against j is $p_{ij}(\gamma_i > \gamma_j) = \frac{e^{\beta_i}}{e^{\beta_i} + e^{\beta_j}}$. The relative strength of i vs. j is given by $\log\left(\frac{p_{ij}}{1-p_{ij}}\right) = \beta_i - \beta_j$. To make the model identifiable one of the strategy coefficients is fixed to zero and all the remaining coefficients are then interpreted as the relative ability compared to the fixated strategy. We fixate when no PDDR strategy is applied. Hence, $\beta_{noPDDR} = 0$.

5.2   Effects of PDDR Strategies on (Un)Fairness

The results of our simulation study concerning the effect on unfairness are presented in Table 1 (note that we do not report p-values separately as due to the large number of simulated outcomes, they are less interesting. However, they are significant on any conventional significance level except for the outcomes marked with "n.s."). Coefficients have to be interpreted in comparison to the case of no handling data deletion requests. The unfairness criterion is $Unf.^{+}(young, \text{old}) = |1 - TPR(young)/TPR(old)|$, and predictive performance is given by accuracy (acc.). We only included simulation runs in the analysis that started with an unfair situation for the young group.

For the DI fairness criterion, which is not reported in the Table 1, only deviations from the threshold of 0.8 are of concern. However, for virtually all methods, this threshold was never violated. Only for table anonymization in 3% of the cases, DI was below this threshold.

Table 1. BTM estimates of PDDR abilities β_i (standard error in brackets, [ns] p-value > 0.1/non-significant, $n = 1.315$).

Variable setting	Unf. + low	Unf. + high	TPR^{young} low	TPR^{young} high	Acc. low	Acc. high
Record deletion	−0.034 (0.028)	0.036[ns] (0.028)	0.259 (0.029)	0.343 (0.030)	0.336 (0.026)	1.191 (0.03)
Record masking	0.610 (0.029)	0.772 (0.029)	−0.487 (0.030)	−0.637 (0.031)	−0.152 (0.026)	−0.005[ns] (0.027)
Table anonymization	2.902 (0.047)	2.699 (0.043)	−3.374 (0.061)	−3.551 (0.064)	−1.129 (0.028)	−1.874 (0.037)
Attribute deletion	−1.017 (0.031)	−0.952 (0.030)	1.834 (0.037)	1.624 (0.035)	0.349 (0.026)	0.657 (0.028)

Table 2. Overview of hypothesis.

Hypothesis		Description	Result
H1	$\beta_{record\ deletion}^{Unf+} > 0 \beta_{masking}^{Unf+} > 0$	Record deletion and masking are unfair	Partially confirmed
H2	$\exp\left(\beta_{record\ deletion}^{Unf+} - \beta_{masking}^{Unf+}\right) > 1$	Record deletion is more unfair than masking	Not confirmed
H3	$\beta_i^{Unf+,low} < \beta_i^{Unf+,high} 0 > \beta_i^{Acc,low} > \beta_i^{Acc,high}$	More PDDR leads to more unfairness and worse overall predictive accuracy	Partially confirmed
H4	$\beta_{Anonymization}^{Unf+} \approx 0$	Anonymization does not affect unfairness	Not confirmed
H5	$\beta_{Attribute\ deletion}^{Unf+} \leq 0$	Attribute deletion does not increase unfairness	Confirmed

6 Discussion

6.1 Results

This research aims to provide light on the subject of which PDDR technique is best suited to ensure fairness. As far as we know, it is one of the first research studies to conceptualize the strategies that a digital service provider might employ to address PDDR. We also discussed the relationship between those strategies and algorithmic fairness and how those two goals might interact. Our findings (see Table 1) indicate that there is such a relationship concerning that research questions. In particular, H1 is partially confirmed as masking, and but not deletion increases unfairness. H2 is not confirmed as record masking leads to higher unfairness. Our hypothesis that total record loss is always worse than decreasing record informativeness was proven inaccurate here. The masking/imputation procedure impaired the model performance unevenly for the younger group. Therefore, the trade-off between complying with the privacy requirement (and discarding all information about the record) and utility/fairness considerations must be balanced (e.g., utilizing a masking strategy that acknowledges, e.g., the age range).

The comparison between the low and high scenarios shows that most PDDR strategies have higher unfairness in the high setting, confirming H3 except for table anonymization. Contrary to our expectation H4, anonymizing the whole table increased unfairness even though both groups were affected by the k-anonymization procedure. Overall predictive accuracy was also reduced. Differential privacy, a different privacy technique from the one we studied, has also been demonstrated in a different setting to have an adverse effect on fairness [40, 41]. Our hypothesis 5 is also confirmed. Indeed, attribute deletion is the method that can reduce unfairness compared to the benchmark case. In so far, attribute deletion turned out to be, in the studied dataset, the best method under consideration. In general, it is also important to consider the target variable's relationship with the attribute being deleted. Younger had a greater default chance in the german credit data set, and, only without, information discriminating between age and predictions for old and young became more similar. As attributes highly correlated with a sensitive attribute may be present in a dataset, attribute deletion is not always recommendable to address fairness [42]. Many enterprises might discard demographic attributes entirely from the ML system in an attempt to develop neutral ML systems. This corresponds to our setting of 'attribute deletion' with implications, as pointed out before.

6.2 Limitations

In the future, more mechanisms that simulate data deletion requests should be investigated. We only examined the outcome when group-specific data deletion requests occurred. Our methodology is consistent with earlier empirical findings that the distribution of PDDR is sociodemographically disparate [43]. Also, we chose a random mechanism $P(\text{"solicit PDDR"}|x \in g_1) = \tau$ and $P(\text{"solicit PDDR"}|g_2 \in x = 0$. A conditional-random mechanism, $P(PDDR = 1|x \in g_1) = \{\tau \ if \ h(x) = 1 \ and \ 0 \ else\}$, may be possible such that only individuals that did not receive the favorable outcome solicit PDDR. But, many digital services fail to clearly convey to users what the favorable outcome could have been (e.g., displaying a high-profile job advertisement vs. displaying a low-profile job advertisement), and as a result, users may solely suspect if they were discriminated against because they did not see the alternative outcome. Such settings can thus be modeled using our selective-random PDDR mechanism.

The dataset we studied is a well-recognized and well-established dataset in the corresponding research literature [8]. Our results may, however, only apply to the specific situation we studied. Nevertheless, we provide a piece of evidence for the design and choice of PDDR strategies.

We only investigated logistic regression. At the same time, machine learning schemes that are transparent and explainable are often preferred over non-transparent machine learning schemes [44], making a transparent machine learning technique such as logistic regression a sensible default model choice. Therefore, we also focused on logistic regression. However, future research directions may address whether the effects may differ across different machine learning techniques. Finally, we did not study the perception of users towards these PDDR strategies. However, this could be an attractive future research stream.

6.3 Implications for Practice and Research

Several lessons can be drawn from our research. First, different PDDR strategies may lead to different outcomes in fairness and predictive performance, making it important to consider carefully which method may be appropriate. Second, careful implementation of masking procedures deserves some attention as an inferior implementation can also cause unfairness and degrade model performance even though the number of affected records is low. Such a masking procedure should acknowledge the original attribute value of the masked record while remaining anonymous. Finally, practitioners also should consider how PDDR affects the model's performance for the remaining users in the database. In general, such evaluation studies as the ones we described may be used to justify certain PDDR strategies to regulators and national data protection agencies.

References

1. Phelps, J., Nowak, G., Ferrell, E.: Privacy concerns and consumer willingness to provide personal information. J. Public Policy Mark. **19**, 27–41 (2000)
2. Dinev, T., Hart, P.: An extended privacy calculus model for e-commerce transactions. Inf. Syst. Res. **17**, 61–80 (2006)
3. Milne, G.R., Rohm, A.J.: Consumer privacy and name removal across direct marketing channels: Exploring opt-in and opt-out alternatives. J. Public Policy Mark. **19**, 238–249 (2000)
4. Gunarathne, P., Rui, H., Seidmann, A.: Racial discrimination in social media customer service: evidence from a popular microblogging platform. In: Bui, T. (ed.) Proceedings of the 52nd Hawaii International Conference on System Sciences. Proceedings of the Annual Hawaii International Conference on System Sciences. Hawaii International Conference on System Sciences (2019). https://doi.org/10.24251/HICSS.2019.815
5. Hardt, M., Price, E., Srebro, N.: Equality of Opportunity in Supervised Learning. In: Proceedings of the 30th International Conference on Neural Information Processing Systems, pp. 3323–3331. Curran Associates Inc., Barcelona (2016)

6. Feldman, M., Sorelle, A., Friedler, J.M., Scheidegger, C., Venkatasubramanian, S.: Certifying and Removing Disparate Impact. In: Proceedings of the 21th ACM SIGKDD International Conference on Knowledge Discovery and Data Mining, pp. 259–268. Association for Computing Machinery, Sydney (2015). https://doi.org/10.1145/2783258.2783311

7. Barocas, S., Selbst, A.D.: Big data's disparate impact. Calif. Law Rev. **104**, 671–732 (2016)

8. Haas, C.: The price of fairness. a framework to explore trade-offs in algorithmic fairness. In: ICIS 2019 Proceedings, vol. 19 (2019)

9. Sweeney, L.: Discrimination in online ad delivery. Google ads, black names and white names, racial discrimination, and click advertising. ACM Queue **11**, 10–29 (2013). https://doi.org/10.1145/2460276.2460278

10. Villaronga, E.F., Kieseberg, P., Li, T.: Humans forget, machines remember: artificial intelligence and the right to be forgotten. Comput. Law Secur. Rev. **34**, 304–313 (2018). https://doi.org/10.1016/j.clsr.2017.08.007

11. Austrian data protection authority: DSB-D123.270/0009-DSB/2018 (2018)

12. Ministry of Electronics & Information Technology, Government of India: The personal data protection bill (2018). https://www.meity.gov.in/writereaddata/files/Personal_Data_Protection_Bill,2018.pdf

13. California Consumer Privacy Act of 2018 [1798.100 - 1798.199.100] (2018). https://leginfo.legislature.ca.gov/faces/codes_displayText.xhtml?division=3.&part=4.&lawCode=CIV&title=1.81.5

14. Data Protection Act 2018 (2021). https://www.legislation.gov.uk/ukpga/2018/12/contents/enacted

15. Sweeney, L.: Achieving k-anonymity privacy protection using generalization and suppression. Internat. J. Uncertain. Fuzzi. Knowl.-Based Syst. **10**, 571–588 (2002)

16. Sweeney, L.: k-anonymity: a model for protecting privacy. Internat. J. Uncertain. Fuzzi. Knowl.-Based Syst. **10**, 557–570 (2002). https://doi.org/10.1142/S0218488502001648

17. Fredrikson, M., Jha, S., Ristenpart, T.: Model inversion attacks that exploit confidence information and basic countermeasures. In: Proceedings of the 22nd ACM SIGSAC Conference on Computer and Communications Security, pp. 1322–1333 (2015). https://doi.org/10.1145/2810103.2813677

18. Veale, M., Binns, R., Edwards, L.: Algorithms that remember: model inversion attacks and data protection law. Philos. Trans. Act. R. Soc. A, **376,** 0083(2018). https://doi.org/10.1098/RSTA.2018.0083

19. Shokri, R., Stronati, M., Song, C., Shmatikov, V.: Membership inference attacks against machine learning models. In: 2017 IEEE Symposium on Security and Privacy (SP), pp. 3–18 (2017). https://doi.org/10.1109/SP.2017.41

20. Tramèr, F., Zhang, F., Juels, A., Reiter, M.K., Ristenpart, T.: Stealing machine learning models via prediction APIs. In: 25th USENIX Security Symposium (USENIX Security 16), pp. 601–618. USENIX Association, Austin (2016)

21. Speicher, T., et al.: Potential for discrimination in online targeted advertising. In: Sorelle, A., Friedler, C.W. (eds.) Proceedings of the 1st Conference on Fairness, Accountability and Transparency. Proceedings of Machine Learning Research, vol. 81, pp. 5–19. PMLR, New York (2018)

22. Köchling, A., Wehner, M.C.: Discriminated by an algorithm: a systematic review of discrimination and fairness by algorithmic decision-making in the context of HR recruitment and HR development. Bus. Res. **13**(3), 795–848 (2020). https://doi.org/10.1007/s40685-020-00134-w

23. Cao, Y., Yang, J.: Towards making systems forget with machine unlearning. In: 2015 IEEE Symposium on Security and Privacy, pp. 463–480 (2015). https://doi.org/10.1109/SP.2015.35

24. Ginart, A., Guan, M., Valiant, G., Zou, J.Y.: Making AI forget you: data deletion in machine learning advances in neural information processing systems, pp. 3518–3531 (2019)

25. Guo, C., Goldstein, T., Hannun, A., van der Maaten, L.: Certified data removal from machine learning models, pp. 3832–3842 (2020)
26. Kazemi, E., Zadimoghaddam, M., Karbasi, A.: Scalable deletion-robust submodular maximization: data summarization with privacy and fairness constraints. In: International Conference on Machine Learning. pp. 2544–2553 (2018)
27. Fung, B.C.M., Wang, K., Chen, R., Yu, P.S.: Privacy-preserving data publishing. ACM Comput. Surv. **42**, 1–53 (2010). https://doi.org/10.1145/1749603.1749605
28. Mivule, K., Turner, C.: A comparative analysis of data privacy and utility parameter adjustment, using machine learning classification as a gauge. Procedia Comput. Sci. **20**, 414–419 (2013). https://doi.org/10.1016/j.procs.2013.09.295
29. Karr, A.F., Kohnen, C.N., Oganian, A., Reiter, J.P., Sanil, A.P.: A framework for evaluating the utility of data altered to protect confidentiality. Am. Stat. **60**, 224–232 (2006). https://doi.org/10.1198/000313006X124640
30. Domingo-Ferrer, J., Torra, V.: A quantitative comparison of disclosure control methods for microdata. In: Doyle, P., Lane, J.I., Theeuwes, J.J.M., Zayatz, L.V. (eds.) Confidentiality, Disclosure and Data Access: Theory and Practical Applications for Statistical Agencies, pp. 111–133, Elsevier (2001)
31. Chamikara, M., Bertok, P., Liu, D., Camtepe, S., Khalil, I.: Efficient privacy preservation of big data for accurate data mining. Inf. Sci. **527**, 420–443 (2020). https://doi.org/10.1016/j.ins.2019.05.053
32. Soria-Comas, J., Domingo-Ferrer, J.: Differentially private data publishing via optimal univariate microaggregation and record perturbation. Knowl.-Based Syst. **153**, 78–90 (2018). https://doi.org/10.1016/j.knosys.2018.04.027
33. Farshid, S., Reitz, A., Roßbach, P.: Design of a forgetting blockchain: a possible way to accomplish GDPR compatibility. In: Proceedings of the 52nd Hawaii International Conference on System Sciences, pp. 1–9 (2019). https://doi.org/10.24251/HICSS.2019.850
34. Chen, M., Zhang, Z., Wang, T., Backes, M., Humbert, M., Zhang, Y.: When machine unlearning jeopardizes privacy. arXiv: Cryptography and Security (2020)
35. Mullainathan, S., Obermeyer, Z.: Does machine learning automate moral hazard and error? Am. Econ. Rev. **107**, 476–480 (2017). https://doi.org/10.1257/aer.p20171084
36. Ensign, D., Friedler, S.A., Neville, S., Scheidegger, C., Venkatasubramanian, S.: Runaway feedback loops in predictive policing. In: Sorelle, A., Friedler, C.W. (eds.) Proceedings of the 1st Conference on Fairness, Accountability and Transparency, vol. 81, 160–171. PMLR, Proceedings of Machine Learning Research (2018)
37. Sweeney, L.: Datafly: a system for providing anonymity in medical data. In: Lin, T.Y., Qian, S. (eds.) Database Security XI. IFIP Advances in Information and Communication Technology, pp. 356–381. Springer, Boston (1998). https://doi.org/10.1007/978-0-387-35285-5_22
38. Kuhn, M.: Caret: Classification and Regression Training. Astrophysics Source Code Library, ascl:1505.003 (2015)
39. Bradley, R.A., Terry, M.E.: Rank analysis of incomplete block designs: I. The method paired comparisons. Biometrika **39**, 324 (1952). https://doi.org/10.2307/2334029
40. Salas, J., González-Zelaya, V.: Fair-MDAV: an algorithm for fair privacy by microaggregation. In: Torra, V., Narukawa, Y., Nin, J., Agell, N. (eds.) MDAI 2020. LNCS (LNAI), vol. 12256, pp. 286–297. Springer, Cham (2020). https://doi.org/10.1007/978-3-030-57524-3_24
41. Bagdasaryan, E., Poursaeed, O., Shmatikov, V.: Differential privacy has disparate impact on model accuracy. In: Wallach, H., Larochelle, H., Beygelzimer, A., d'Alché-Buc, F., Fox, E., Garnett, R. (eds.) Advances in Neural Information Processing Systems, vol. 32. Curran Associates, Inc (2019)

42. Dwork, C., Hardt, M., Pitassi, T., Reingold, O., Zemel, R.: Fairness through awareness. In: Proceedings of the 3rd Innovations in Theoretical Computer Science Conference, pp. 214–226. Association for Computing Machinery, Cambridge (2012). https://doi.org/10.1145/209 0236.2090255

43. Johnson, G.A., Shriver, S.K., Du, S.: Consumer privacy choice in online advertising: who opts out and at what cost to industry? Mark. Sci. **39**, 33–51 (2020). https://doi.org/10.1287/mksc.2019.1198

44. Wang, R., Harper, F.M., Zhu, H.: Factors influencing perceived fairness in algorithmic decision-making. In: Bernhaupt, R. (ed.) Proceedings of the 2020 CHI Conference on Human Factors in Computing Systems. ACM Digital Library, pp. 1–14. Association for Computing Machinery, New York, NY, United States (2020). https://doi.org/10.1145/3313831.3376813

Empirical Investigations on Digital Innovation and Transformation Prerequisites

How Companies Develop a Culture for Digital Innovation: A Multiple-Case Study

Patrizia Orth[1]([✉]) [iD], Gunther Piller[1] [iD], and Franz Rothlauf[2] [iD]

[1] University of Applied Sciences Mainz, Mainz, Germany
{patrizia.orth,gunther.piller}@hs-mainz.de
[2] Johannes Gutenberg University Mainz, Mainz, Germany
rothlauf@uni-mainz.de

Abstract. Companies are constantly facing new challenges through digital technologies, as they can cause rapid changes in their business environment. Digital innovations can offer the opportunity to gain competitive advantages, but in practice, companies often struggle to create the conditions for this. Developing a culture for digital innovation is an important capability that companies need to build. It includes appropriate organizational structures, working methods, skills, and new ways of thinking in the innovation process. This has been explored in previous research at a high level of abstraction. Our paper complements past findings with a detailed study of what activities companies are doing in practice and what methods they are using to support these activities to pave the way for digital innovation in their organization. For this purpose, a multiple-case study has been conducted. As a theoretical basis, we use the dynamic capability approach. We obtain how the capability for developing a digital culture is substantiated by microfoundations and underlying activities.

Keywords: Digital innovation · Innovation process · Dynamic capabilities

1 Introduction

Innovations in digital technologies pose challenges for companies, as they can cause disruptive effects on business environments, forcing companies to identify and evaluate new developments at an early stage [1]. However, digital technologies also offer companies opportunities to gain a competitive advantage, e.g., by innovating products, optimizing processes, or opening up new markets or distribution channels [2].

Digital innovation affect companies as a whole [3]. The ability to innovate is, e.g., influenced by organizational structures [4] and working methods [5]. Digital innovation also requires new skills in the company, such as continuous learning [6] and new ways of how employees think and act [7, 8]. It is therefore important to develop a culture for digital innovation, i.e. to shape the interplay of structures, working methods, skills, and mindsets for innovations based on digital technologies.

The ability of organizations to adopt and generate value from digital technologies is often studied from the perspective of dynamic capabilities. Here, one generally focuses

© Springer Nature Switzerland AG 2021
R. A. Buchmann et al. (Eds.): BIR 2021, LNBIP 430, pp. 221–235, 2021.
https://doi.org/10.1007/978-3-030-87205-2_15

on an organization's ability to create, change, or renew resources or its mix in response
to a rapidly changing environment [9, 10]. The dynamic capabilities are substantiated
by microfoundations consisting of individual and group level activities [11].

In the literature, the adoption of digital innovations using the capability approach
is discussed in various domains, see e.g., [12–16] and abilities regarding organizational
structures, working methods, skills, and mindsets have been identified as fundamental,
see e.g., [17–19]. Dynamic capabilities are a well-accepted concept in the context of
digital innovation. However, how the corresponding microfoundations are concretized
has not yet been examined in detail. As, e.g., Vial [16] noted, it would be important "to
further engage with the nature of the work performed by actors" that support dynamic
capabilities and Verhoef et al. [20] identified research opportunities asking questions
such as "How can firms develop specific digital resources?" such as digital capabilities.

We want to fill this gap. Our paper aims to extend existing research on dynamic
capabilities for digital innovation by focusing on the capability for developing a digital
culture, with emphasis on its microfoundations and the activities that underpin them.
From a process perspective, the initiation phase of digital innovation, where the potentials
of new technologies are explored [21, 22] is particularly interesting as it represents a
serious first hurdle for technology-driven innovation. Since digital culture accompanies
the entire innovation process, this first step is also the focus of our study.

Starting from the dynamic capability and the higher level of abstraction of *what*
companies do to pave the way for digital innovation in the organization, we specifically
address the question *how* they do it, what activities companies are doing in practice and
what methods they use to support their activities. For this purpose, we follow a qualitative
approach, which was implemented in a two-stage process. First, the microfoundations
of dynamic capabilities for digital innovation from the literature were examined with
respect to their relation to digital culture and consolidated inductively. After that, in a
deductive phase, a multiple-case study was conducted with companies that have been
active in digitalization projects for several years and have implemented initiatives to
drive innovation through digitalization within their organizations.

The theoretical underpinnings are discussed in Sect. 2 and 3 before the research
design is presented in Sect. 4. A multiple-case study (see Sects. 5–7) demonstrates how
companies foster a digital culture in practice. We summarize in Sect. 8.

2   Dynamic Capabilities and Innovation

From a theoretical perspective, we use the concept of dynamic capabilities [10] and the
innovation process in organizations [22] as a framework to structure our study and discuss
its results. According to Rogers [22], the innovation process consists of two phases
"initiation" and "implementation", separated by the decision to adopt an innovation. In
the initiation phase, companies decide how to react to the availability of a new technology,
identify problems and needs, scan the business environment for potentially beneficial
innovations, and finally examine whether an innovation is suitable to meet the needs of
the organization. At the end, a decision for adoption is made and – in case of a positive
result – the implementation phase is initiated. Since the initiation phase forms the basis
for innovation decisions and represents a first severe hurdle for many companies [21],
our research focuses on this stage.

Rogers's innovation process [22] describes on a very general level how companies explore and adopt innovations. On the other hand, dynamic capabilities play a major role for companies facing technological innovations [18, 19] and have become the default answer to what enables firms to deal with technological change [21]. Dynamic capabilities emerged from a resource-based view of a firm to explain improvements in competitive advantage as a result of changing value-creating resources over time [7, 10]. Dynamic capabilities enable businesses to build, extent, modify, and reconfigure tangible and intangible assets, the products and services offered, as well as customers and ecosystems. In the initiation phase of the innovation process, two types of dynamic capabilities are of crucial importance. The sensing capability with its ability to recognize emerging developments and shape new opportunities [16, 19, 21, 23, 24] and the dynamic capability to develop an appropriate culture that accompanies the innovation process from the beginning, e.g., through the ability to create organizational knowledge and enable new ways of thinking [3, 7, 8]. Two other dynamic capabilities are described in the literature: "seizing", i.e., implementing identified opportunities, and "transforming", i.e., aligning and adapting the firm's assets to remain profitable in the long term [24]. However, since these capabilities are most relevant for the later phases of the innovation process [21], they are not the focus of our work.

3   Review and Consolidation of Microfoundations

Previous research has revealed various microfoundations related to the capability for developing a digital culture. To obtain a common ground for our research, we have compared and consolidated these microfoundations.

In order to identify core papers, we searched for relevant publications on dynamic capabilities in connection with innovation and digital innovation. To determine whether the publications were relevant, we manually worked through the search results. We excluded publications that discussed dynamic capabilities at a very high a level of abstraction without considering the underlying microfoundations. Our analysis considered studies regarding dynamic capabilities, e.g., for sustained value creation [17], for profiting from innovation [12], in responding to digital disruption [18], for digital transformation [8, 14, 16], to renew information technology (IT)-based resources [26], for innovation networks [15], to sustain competitive advantage in economies with rapid innovation [24, 27], for leveraging big data [25], and to enact a digital strategy [19].

The microfoundations identified in the literature fall into two categories. The first category covers activities related to skills at organizational and individual level. This includes training and knowledge sharing [14] and recruiting "people with specific skill sets to complement the competence base" [17]. Beyond that, awareness of how knowledge can be generated from data [19, 25] also helps firms build up digital qualifications. The second category summarizes activities related to developing and optimizing a digital organization. This includes, e.g., establishing cross-functional teams [8], but also "encouraging the exploration of new ideas" [17] and communicating "a clear, albeit broad, [digital] strategy to all levels of the organization" [19].

We analyzed suggestions for developing a digital culture (see the "Key findings" column of Tables 1 and 2) and consolidated the microfoundations of the two categories into the microfoundations for "Building and Maintaining Digital Qualifications" (see Table 1) and "Developing and Optimizing a Digital Organization" (see Table 2).

Table 1. Consolidated microfoundation: building and maintaining digital qualifications

Authors	Key findings
Achtenhagen et al. [17]	"Hiring or cooperating with people with specific skill sets to complement the competence base"
Helfat and Raubitschek [12]	"As part of this process [...] may need to develop or acquire new skills, knowledge [...]"
Karimi and Walter [18]	"[...] attracting and retaining digital talent in a traditional culture can be very difficult."
Li et al. [14]	"Training team members [...]", "Knowledge sharing [...]", "filling key [...] positions"
Rehm et al. [15]	"Providing systematic processes and tools for synchronizing information and exchanging knowledge"
Teece [24]	"The combination of know-how within the enterprise, and between the enterprise and organizations external to it (e.g., other enterprises, universities), is important."
Teece et al. [27]	"Knowledge and capabilities are not only scarce but also often difficult to imitate. Sometimes they can be bought; generally, they have to be built."
Tiefenbacher and Olbrich [25]	"the capability [...] needs to be further developed with regard to new skills and expertise"
Vial [16]	digital transformation raises the question of developing the skills of existing workforce and the required skills of the future digital workforce
Warner and Wäger [8]	"improving the digital maturity of the workforce"
Yeow et al. [19]	"learning new competencies", "hiring of people"

A systematic investigation of the activities through which these microfoundations are realized in practice has not yet been carried out. We close this gap by the present work. The microfoundations consolidated in this section from previous work form the starting point for this.

Table 2. Consolidated microfoundation: developing and optimizing a digital organization

Authors	Key findings
Achtenhagen et al. [17]	"Providing freedom for and encouraging the exploration of new ideas", "Focusing on open communication [...]"
Karimi and Walter [18]	"To successfully manage innovation projects, then, requires firms to build and sustain an innovation-supportive culture [...]", "It [pursuing a digital strategy] will require developing a digital-first mindset, making digital strategy everyone's job, cultivating an open culture that embraces digital innovations [...]"
Queiroz et al. [26]	The ability of firms to renew IT-based resources improves agility and performance
Rehm et al. [15]	"Establishing operative working infrastructures and team spirit"
Teece [24]	"entrepreneurial managerial capitalism [...] involves [...] reshaping organizational structures and systems so that they create and address technological opportunities [...]"
Teece et al. [27]	"It has long been recognized that although costly, "slack" (maintaining excess resources and capacity) assists agility [...] Organizational structure also has strong implications for agility."
Vial [16]	"In the context of DT [digital transformation], organizational leaders must work to ensure that their organizations develop a digital mindset [...]", "In the context of DT [digital transformation], changes to the structure as well as the culture of an organization lead employees to assume roles that were traditionally outside of their functions."
Warner and Wäger [8]	"Enabling an entrepreneurial mindset", "Promoting a digital mindset", using "Cross-functional teams"
Yeow et al. [19]	"Communicate a clear, albeit broad, [digital] strategy to all levels of the organization", "invest in creating an entirely new [digital] department", "leverage the existing workforce in a cross departmental manner"

4 Research Design

Starting with our research questions and the theoretical lens from the previous section, which combines Rogers's innovation process [22] with the dynamic capability for developing a digital culture and corresponding consolidated microfoundations from the literature, we entered the second stage of our research. Since we address typical "how" questions to obtain an in-depth understanding of contemporary and complex phenomena in a real-life context, the case study approach in the second part of our investigation is appropriate [28, 29]. We especially payed attention to the four established quality criteria: external validity, internal validity, construct validity, and reliability [29].

To increase the generalizability of our results, we use a multiple-case design [28, 29]. The data collection started with the development of an interview guideline. Then case selection and in-depth interviews followed. We also collected additional data, like project reports or publicly available information. The subsequent data analysis included examination of the interviews and secondary data, a triangulation, and a final cross-case analysis. As a result, we obtained detailed insights on empirically validated activities that show how to support the development of a digital culture that accompanies digital innovations in the beginning of the innovation process.

5 Data Collection

Interview Guideline

In order to ensure internal validity, the interview guideline was based on the consolidated microfoundations from Sect. 3. We used a semi-structured questionnaire for each interview. This ensured that all relevant aspects were covered, but also provided a flexible response to the informant's answers [30]. The interview guideline consisted of a section with general questions, e.g., about the company, the interviewee, their area of responsibility, and the corresponding organizational setup. Then questions followed about workforce development, changes in organizational structures, and adjustments to working methods in relation to digital technologies. We formulated open questions so that respondents could answer in their own words instead of limiting themselves to predetermined or preferred terms or phrases [30]. We consulted an expert with years of hands-on experience in technology management to get external feedback on the interview guideline. After making minor adjustments, a pilot interview was conducted with a head of a business development department to cover also the business perspective.

Case Selection

To ensure external validity, which focuses on the generalizability of the results, we followed the literal replication logic by selecting cases with a similar contextual background [29]. The selected cases met the following criteria. First, to gain valuable insights into existing approaches to digital innovation activities, we selected companies that have been active in digital innovation projects for several years. Furthermore, we chose companies that were similar in the following aspects: large enterprises with more than 10.000 employees, companies that operate internationally and have their headquarters in Germany, and companies that are active in business areas of the manufacturing industry.

This constraints fit also well to our focus on companies being active in digital transformation, since in general, large companies are expected to be less constrained in their access to resources and capabilities for innovation than small enterprises [31]. Moreover, companies that operate internationally are typically exposed to a diverse, complex, and dynamic market environment and must therefore maintain their competitiveness through innovation and flexibility [32]. The manufacturing industry also fits well into our research, as it has made considerable progress in digitalization in Germany in recent years, while digitalization activities in other economic sectors have often remained at the same level [33]. It was also important for us to address key informants at a high decision-making level, which is why we chose companies with headquarters in Germany. Interviewees were decision-makers responsible for digital technology adoption and transformation programs and related innovation initiatives, such as Chief Digital Officers (CDO) and Heads of Innovation & Digitalization.

For the purposes of obtaining an external view, we also included consulting firms in our research. Here, we interviewed consultants responsible for projects on digital technology strategies and large-scale transformation efforts in companies that met the selection criteria described above. The consultants were at senior and partner level and were members of five management, strategy and technology consulting firms.

In order to identify the appropriate interviewees, various information was gathered in advance about the companies and consultancies, their digitalization projects, as well as the eligible key informants and their background. The respondents were recruited through invitation emails or by contacting them via professional social networks. Of 58 people contacted from 39 companies, 10 agreed to be interviewed. Table 3 provides an overview of the investigated cases and involved interview partners.

Table 3. Overview of sample properties

Firm	Industry sector	Headcount	Interviewee position
A	Chemicals	>20,000	CDO
B	Energy	>20,000	Corporate Head Digitalization
C	Mechanical engineering	>12,000	Manager Digital Innovation Unit
D	Consumer goods	>50,000	Corporate Director Data & Application
E	Pharmaceutical	>50,000	Manager IT Strategy
F	Consulting	>20,000	Partner, technology transformation
G	Consulting	>20,000	Partner, technology transformation
H	Consulting	>100,000	Director, technology transformation
I	Consulting	>100,000	Consultant, technology transformation
J	Consulting	>400	Partner, technology transformation

Interviews and Additional Data

We conducted the interviews over a period of five months. The participants were informed in advance about the content of the interviews and we guaranteed the confidentiality of the responses. The interviews lasted approximately one and a half hours.

With the aim of enabling data triangulation and improving construct validity, we used also additional data sources [29]. We conducted desk research to collect supplementary secondary data, e.g., freely available reports, white papers, presentations, and up-to-date news on each company's digitalization and innovation initiatives.

In order to avoid falsifications and ensure reliability, all cases were organized in the same way: all conversations were digitally recorded and shortly after the interview transcribed; all data was stored in a case study database [29].

6   Data Analysis

In the early steps of the data analysis, the main topics and main statements of the interviews were summarized for each case with the field notes of the interviewers in a short report. This was done in order to get a quick overview without losing the essential information of the interviews [34].

A directed content analysis approach was adopted [35]. The data analysis began by reviewing all interviews and coding all text passages that apparently described activities related to the consolidated microfoundations. Codes derived from prior theorizing, especially terms from the preparatory work of consolidated microfoundations and the interview template, were used as a pre-set coding scheme [34]. Whenever necessary, codes were added during data analysis to meet the terms used by the informants [30].

For each case, we applied a within-case analysis to identify initial patterns. By triangulating the multiple data for each case, we ensured that the findings came from more than one source of evidence, but still different dimensions of the same phenomenon were captured [29]. This resulted in a collection of detailed information on activities for developing a digital culture and corresponding organizational measures. By merging similar patterns, the identified activities were condensed into six key activities.

An iterative approach was chosen, by repeatedly comparing the analysis results with the consolidated microfoundations from the literature [30]. An initial understanding of the relationships between the different microfoundations and the related activities was a key result here.

We completed the data analysis with the cross-case analysis. To compare the cases to each other contributes to the generalizability [28] and enhances understanding and explanation [34] of the results by discerning consistent patterns and differences between cases. As a final result, we obtained a consolidated picture about microfoundations and corresponding activities for developing a digital culture.

7   Results

In our cases, all activities could be related to the microfoundations from theorizing in Sect. 3. The observed activities manifesting the microfoundations *"Building and Maintaining Digital Qualifications"* and *"Developing and Optimizing a Digital Organization"* and thus underpinning the dynamic capability for developing a digital culture are described in detail below. With regard to all key activities contributing here, it was emphasized that they are essential for the initiation phase of digital innovation, but they are not one-off activities. They are to be understood as continuous tasks that are necessary to refresh skills and competences or to adapt them to changing requirements. It was also pointed out, that these activities need to be rolled out throughout the organization, covering all functional areas and hierarchies.

Additionally, in examining the identified key activities, we found that they are composed of interrelated elements of procedures, processes, structures, and systems, which Teece [24] cited as examples of the composition of the microfoundations. From these findings, a structure of the activities results, as shown in Tables 4 and 5. It is interesting to note that all observed key activities cross organizational boundaries, either within or between companies.

7.1 Microfoundation "Building and Maintaining Digital Qualifications"

The first key activity that was emphasized in all interviews is *"Promote Use of Digital Technology"*. Employees must be prepared for the changes brought about by digitalization. Training the workforce in the use of digital systems and actively promoting their usage is therefore essential. Trainings must be provided in a well-organized way on different levels for all organizational areas of the company and all types of systems, like communication tools, or systems to support specific business processes, e.g., customer relationship management (CRM) systems, or enterprise resource management (ERP) systems. Training for specialist systems is also required, e.g., for the usage of analysis platforms by data scientists. In our cases, it was pointed out that the training processes need to be tailored to different target groups and different methods and forms are applied to train employees. They range from, e.g., classroom sessions to opportunities to participate in web-based introductory courses that can be taken on an individual basis. The ability to connect with others and build knowledge networks through internal communities was seen as an important aspect to promote the use of new systems and technologies. One respondent (Case E) stated in the interview:

> *"We [Department for IT Strategy] thought about how we could help employees to build up or expand the necessary skills through training, or by forming networks where people could exchange ideas. [...] There are also series of events where people talk about appropriate topics, where topics are presented and are supposed to stimulate thinking. There are training programs for all kinds of methods, whether it's design thinking, whether it's scrum, so there's a large catalog of offerings."*

The training measures can be supported by corresponding systems, such as platforms for self-paced online trainings. In the following, a digital innovation manager (Case C) explained their web-based training platform for employees:

> *"There are different categories of topics, there is something like 'basics of digitalization', but there are also very, very specific topics for employees that maybe not everyone needs, but that everyone can look at [...]. Of course, employees can also use the platform to exchange information, ask questions and so on, because we always have an expert on every topic."*

In order to motivate employees to engage with the content, this platform also uses gamification approaches. The more online training employees complete and the more content they generate themselves on the platform, the higher the level they hold on the platform. To track their own progress and to compete with others, the different levels are rewarded with virtual colored belts, similar to the ranking scheme in judo or karate.

To leverage digital innovations, companies have to be able to derive insights from data, i.e., they need to know how to use data to generate value. In the case of customer analytics using big data technologies, this has been pointed out, e.g., by Tiefenbacher and Olbrich [25]. This ability is significantly promoted by the key activity *"Enable Data Thinking"*, which was highlighted in our interviews (Cases A, B, D, F, H). It fosters awareness of opportunities to create value from data. Continuous training of employees from all departments on the potential of data for processes, products, and services is seen by those surveyed as a central element. Method training on ways to identify new, data-driven scenarios, such as design thinking, was also mentioned as an important component. Efforts to ensure data transparency so that everyone knows what data is available in the organization were considered mandatory, as well as data availability using, e.g., centralized databases or data platforms, and data analytics using, e.g., big data systems and advanced analytics tools (Cases D, F, H). In most cases, these activities were coordinated by units responsible for innovation or digitalization, such as a digital center of excellence. A consultant (Case H) emphasized in this context:

> "Another topic is certainly data [...], so that you also try to generate added value from all the data solutions. So how you can use them [...]."

The activity "Enable Data Thinking" also involves promoting awareness among employees and managers that data is a valuable asset and should be collected and handled with high care to ensure availability and quality. A related aspect added by one respondent (Case D) was data accountability:

> "We're working on a dedicated data strategy [...]. I'm a fan of incentivizing that and really holding our managers accountable not only for the processes but also for the data that comes out of it, because that's a huge contribution to basically managing the whole company digitally, but that's a big culture change that's still ahead of us."

To create sustainable value in exploiting new business opportunities, it could be necessary to hire people with specific skills, as observed by, e.g., Achtenhagen et al. [17]. "Recruit Digital Talents" is the related key activity confirmed in almost all cases (Cases A, B, C, D, E, F, H, I). A job analysis helps to determine the knowledge and skills associated with jobs and derive personnel requirements from that, as one interviewee (Case E) noted. In our cases, recruitment was necessary to add digital expertise to specialist departments. For example, in Case C, to build a digital sales force, the company assembled a team consisting of company sales representatives who brought valuable expertise about the products and the company and supplemented this team with new employees who didn't know the industry but were knowledgeable about digital sales or digital marketing. Recruitment is also used to fill positions in newly established organizational units, e.g., data analytics units or digital innovation teams. The corporate head digitalization (Case B) remarked on this:

> "You are talking to someone [who has been hired for digitalization initiatives himself] and who has hired a lot of people from that direction. So the answer is 'yes', we have hired many, many people to compensate for the lack of expertise."

It was emphasized in our cases that this activity has to be set up cross-functionally. To be successful, human resources (HR) management must work closely with both the departments seeking new employees and the units responsible for digital innovation.

Table 4. Microfoundation "building and maintaining digital qualifications": key activities and examples of characteristic elements from the cases

Key activity	Procedures	Processes	Structures	Systems
Promote Use of Digital Technology – all cases – (cross-functional)	Web-based Introductory Courses	Training	Internal Communities	Web-based Training Platforms
	Classroom Sessions	Learning Journey	Internal Mentoring	Collaboration Platforms
Enable Data Thinking – 5 cases – (cross-functional)	Lean Inception-Workshops	Data Management	Digital Innovation Units	Big Data - & AI-Systems
	Design Thinking	Prototyping	Data Science Labs	Advanced Analytics-Tools
Recruit Digital Talents – 8 cases – (cross-functional)	Job Analysis	Recruitment	HR Departments, Digital Innovation Units	Online Career Portals

7.2 Microfoundation "Developing and Optimizing a Digital Organization"

To operationalize the individual skills and knowledge of employees and orchestrate activities around digital innovations, companies need corresponding structures at the organizational level. Developing and optimizing a digital organization includes activities for creating space for innovative thinking and intrapreneurship, setting up flexible organizational structures, and introducing and advancing agile working methods. Our cases confirm these activities mentioned, e.g., in [8, 15, 27] and add interesting aspects regarding their concrete design. Similar to the activities in the previous section, the importance of engaging employees from all parts of the organization and establishing networks, even across organizational boundaries, was emphasized here too.

"*Create Space for Intrapreneurship*" was mentioned by all cases as an important key activity. It encourages innovative thinking and entrepreneurial action by employees, facilitating the identification and adoption of digital innovations. Similar observations have also been made by Li et al. [36] at small and medium-sized enterprises. In most cases, our interviews showed that it is particularly promising to actively involve employees from different areas of the company in order to obtain information about trends in digital technologies and innovative approaches to their use (Cases B, C, D, E, F, H, I). Some companies support this with innovation platforms where employees can share their ideas or view information on ongoing projects (Cases C, D, E). One interviewee (Case D) described employees as the largest sensory system available to the organization for relaying information, much like a large skin surface that passes perceptions to the brain. On employee engagement and innovative thinking, a consultant (Case I) stated:

> "It is very important to establish the appropriate mindset among employees as the new standard, for example, that things can change. You have to reach a 'tipping point' so that as many employees as possible are taken along and at some point the others also swim with you."

A consultant (Case G) reported on a successful approach to promoting intrapreneurship, where there is an internal innovation incubator with its own budget for digital innovation projects. Departments can apply there and pitch their innovation ideas. Board members involved in the incubator decide which pitch was the most promising and thus which project will be funded and provided with a team.

Also here, our interviewees emphasized the importance of cross-functional networks to foster innovative thinking and intrapreneurship: communities of practice, in which all interested employees can exchange ideas on digital topics and drive digital innovation projects, but also specialized communities for IT professionals, e.g., for data scientists (Case B, C, F). A consultant (Case F) reported on this from his projects:

"Many companies are creating communities for their digital talents, such as data scientists, because they have realized that it is not enough for individual IT professionals to be embedded somewhere in the department; they simply won't be happy there. Today's digital talents have different requirements, need their own community, and need to be among their peers to be satisfied and stay with the company longer."

One interviewee (Case D) also pointed out a possible area of tension in practice: If processes in the company are so streamlined that the focus is only on daily work, the employees lack the freedom to additionally deal with digital innovations. This is where organizational approaches to slack time for innovations may help, as discussed by Agrawal et al. [37], for example.

Creating space for intrapreneurship also includes a fault tolerance for innovation activities and welcoming corporate communication on digital initiatives with encouragement for employees to participate (Case B, D, G, H, I). The examples from our cases are recurring activities, ranging from regular newsletters, blogs on the intranet, employee surveys ("pulse checks"), monthly live events where management reports on current developments and employees can ask questions about them, to "open-door sessions" where employees can seek out conversations with C-level executives.

Organizational structures have a direct influence on the ability to innovate, as, e.g., discussed in [4]. Consistently with this, the key activity *"Set up Appropriate Structures"* was mentioned by all our cases. It involves the creation of organizational units responsible for digital innovations, the so-called innovation labs, digital accelerators, or data science labs, just to name a few examples from the interviews. In our cases, various tasks of these organizational units were mentioned, e.g., prioritizing business areas for digital innovations, generating innovation ideas, designing new products, services, or processes, and enabling data thinking. There are various ways to connect innovation units into organizational structures, ranging from purely internal structures to external networks, see e.g., [15]. Involvement in external networks was also mentioned in our interviews (Cases A, B, C, D, F), e.g., in industry associations, start-up networks, research networks with universities, or innovation networks with suppliers and trading partners. External networks may be useful for companies in various digital innovation activities, examples from our cases are identifying technology trends and new applications, exploring market changes, obtaining customer insights, generating innovation ideas, and developing and testing prototypes.

"Introduce Agile Working" is the third key activity for the microfoundation discussed in this section. The general importance of agile methods for innovation has been pointed out in different contexts in, e.g., [8, 13, 36] and we found this topic in our cases too. All of our respondents emphasized the benefits of agile working methods like scrum in digitalization projects, e.g., speed and flexibility. The value of cross-functional teams in agile projects for digital innovations was also stressed by our cases. A consultant (Case F) explained for example:

> *"I'm a big supporter and friend of cross-functional teams. Collaboration, meaning truly agile working, certainly has advantages."*

Some typical activities from our cases that benefit from cross-functional teams are, e.g., contextualizing collected information on digital technologies with business criteria, generating innovation ideas, and designing new products, services, or processes.

In addition, the empowerment of employees, e.g., through the promotion of competencies and responsibilities, appropriate employee roles and corporate hierarchies, was seen as an important factor for the successful use of agile methods (Cases B, E, F, G, H, I), confirming, e.g., the discussion about the adaptability of innovation teams in Grass et al. [38]. However, agile methods are not mandatory in all digital innovation projects. Our respondents pointed to the need to carefully consider which problems can be addressed with agile approaches or where agile working methods should be adapted or replaced by other practices (Cases C, E, F, H, J).

Table 5. Microfoundation "developing and optimizing a digital organization": key activities and examples of characteristic elements from the cases

Key activity	Procedures	Processes	Structures	Systems
Create Space for Intrapreneurship – all cases – (cross-functional)	Pitching Contests	Concept Screening	Internal Incubators	Web-based Suggestion Schemes
	Business Model Canvas	Idea Generation	Communities of Practice	Intranets
Set up Appropriate Structures – all cases – (cross-functional and cross-organizational)	Agile Working Methods	Innovation Management	Digital Innovation Units	Task Management Tools
	Focus Groups	Open Innovation	Customer Network	Idea Collaboration-Tools
Introduce Agile Working – all cases – (cross-functional)	Scrum	Empowerment	Cross-functional Teams	Collaboration Platforms

8 Summary

We have examined how companies pave the way for digital innovation in the initiation phase of the innovation process by developing a digital culture. We consolidated microfoundations of dynamic innovation capabilities from the literature as the basis for a multiple-case study. Our study identified and illustrated six key activities and their interrelated elements of procedures, processes, structures, and systems. All key activities cross organizational boundaries, either within or between companies. Our research synthesized theoretical and empirical findings related to the dynamic capability for developing a digital culture. In addition, our research helps companies align their current practices with the identified approaches. So far, we have only considered activities for developing a digital culture. Activities for other dynamic capabilities related to the initiation phase of the innovation process are the subject of ongoing research.

References

1. Christensen, C.M.: The Innovator's Dilemma: When New Technologies Cause Great Firms to Fail. Harvard Business School Press, Boston (1997)
2. Yoo, Y., Henfridsson, O., Lyytinen, K.: Research commentary—the new organizing logic of digital innovation: an agenda for information systems research. Inf. Syst. Res. 21(4), 724–735 (2010)
3. Kohli, R., Melville, N.P.: Digital innovation: a review and synthesis. Inf. Syst. J. 29(1), 200–223 (2019)
4. Tushman, M., Smith, W.K., Wood, R.C., Westerman, G., O'Reilly, C.: Organizational designs and innovation streams. Ind. Corp. Change 19(5), 1331–1366 (2010)
5. Teece, D.J.: Dynamic capabilities: routines versus entrepreneurial action. J. Manage. Stud. 49(8), 1395–1401 (2012)
6. Nylén, D., Holmström, J.: Digital innovation strategy: a framework for diagnosing and improving digital product and service innovation. Bus. Horiz. 58(1), 57–67 (2015)
7. Eisenhardt, K.M., Martin, J.A.: Dynamic capabilities: what are they? Strateg. Manag. J. 21(10–11), 1105–1121 (2000)
8. Warner, K.S., Wäger, M.: Building dynamic capabilities for digital transformation: an ongoing process of strategic renewal. Long Range Plan. 52(3), 326–349 (2019)
9. Ambrosini, V., Bowman, C.: What are dynamic capabilities and are they a useful construct in strategic management? Int. J. Manage. Rev. 11(1), 29–49 (2009)
10. Teece, D.J., Pisano, G., Shuen, A.: Dynamic capabilities and strategic management. Strateg. Manag. J. 18(7), 509–533 (1997)
11. Eisenhardt, K.M., Furr, N.R., Bingham, C.B.: Microfoundations of performance: balancing efficiency and flexibility in dynamic environments. Organ. Sci. 21(6), 1263–1273 (2010)
12. Helfat, C.E., Raubitschek, R.S.: Dynamic and integrative capabilities for profiting from innovation in digital platform-based ecosystems. Res. Policy 47(8), 1391–1399 (2018)
13. Kranz, J.J., Hanelt, A., Kolbe, L.M.: Understanding the influence of absorptive capacity and ambidexterity on the process of business model change–the case of on-premise and cloud-computing software. Inf. Syst. J. 26(5), 477–517 (2016)
14. Li, L., Su, F., Zhang, W., Mao, J.Y.: Digital transformation by SME entrepreneurs: a capability perspective. Inf. Syst. J. 28(6), 1129–1157 (2018)
15. Rehm, S.V., Goel, L., Junglas, I.: Using information systems in innovation networks: uncovering network resources. J. Assoc. Inf. Syst. 18(8), 577–604 (2017)

16. Vial, G.: Understanding digital transformation: a review and a research agenda. J. Strateg. Inf. Syst. **28**(2), 118–144 (2019)
17. Achtenhagen, L., Melin, L., Naldi, L.: Dynamics of business models–strategizing, critical capabilities and activities for sustained value creation. Long Range Plan. **46**(6), 427–442 (2013)
18. Karimi, J., Walter, Z.: The role of dynamic capabilities in responding to digital disruption: a factor-based study of the newspaper industry. J. Manage. Inf. Syst. **32**(1), 39–81 (2015)
19. Yeow, A., Soh, C., Hansen, R.: Aligning with new digital strategy: a dynamic capabilities approach. J. Strateg. Inf. Syst. **27**(1), 43–58 (2017)
20. Verhoef, P.C., et al.: Digital transformation: a multidisciplinary reflection and research agenda. J. Bus. Res. **122**, 889–901 (2021)
21. Konlechner, S., Müller, B., Güttel, W.H.: A dynamic capabilities perspective on managing technological change: a review, framework and research agenda. Int. J. Technol. Manage. **76**(3–4), 188–213 (2018)
22. Rogers, E.M.: Diffusion of Innovations. Free Press, New York (2003)
23. Teece, D.J.: Towards a capability theory of (innovating) firms: implications for management and policy. Camb. J. Econ. **41**(3), 693–720 (2017)
24. Teece, D.J.: Explicating dynamic capabilities: the nature and microfoundations of (sustainable) enterprise performance. Strateg. Manag. J. **28**(13), 1319–1350 (2007)
25. Tiefenbacher, K., Olbrich, S.: Developing a deeper understanding of digitally empowered customers–a capability transformation framework in the domain of customer relationship management. In: Pacific Asia Conference on Information Systems (PACIS). Association for Information System (2016)
26. Queiroz, M., Tallon, P.P., Sharma, R., Coltman, T.: The role of IT application orchestration capability in improving agility and performance. J. Strateg. Inf. Syst. **27**(1), 4–21 (2018)
27. Teece, D.J., Peteraf, M., Leih, S.: Dynamic capabilities and organizational agility: risk, uncertainty, and strategy in the innovation economy. Calif. Manage. Rev. **58**(4), 13–35 (2016)
28. Benbasat, I., Goldstein, D.K., Mead, M.: The case research strategy in studies of information systems. MIS Q. **11**, 369–386 (1987)
29. Yin, R.K.: Case Study Research and Applications: Design and Methods, 6th edn. Sage, Los Angeles (2017)
30. Gioia, D.A., Corley, K.G., Hamilton, A.L.: Seeking qualitative rigor in inductive research: notes on the Gioia methodology. Organ. Res. Methods **16**(1), 15–31 (2013)
31. Hewitt-Dundas, N.: Resource and capability constraints to innovation in small and large plants. Small Bus. Econ. **26**(3), 257–277 (2006)
32. Zahra, S.A., Ireland, R.D., Hitt, M.A.: International expansion by new venture firms: international diversity, mode of market entry, technological learning, and performance. Acad. Manag. J. **43**(5), 925–950 (2000)
33. Weber, T., Bertschek, I., Ohnemus, J., Ebert, M.: DIGITAL Economy Monitoring Report 2018. Federal Ministry for Economic Affairs and Energy (BMWi) (2018)
34. Miles, M.B., Huberman, A.M.: Qualitative Data Analysis: An Expanded Sourcebook, 2nd edn. Sage, Thousand Oaks (1994)
35. Hsieh, H.F., Shannon, S.E.: Three approaches to qualitative content analysis. Qual. Health Res. **15**(9), 1277–1288 (2005)
36. Li, W., Liu, K., Belitski, M., Ghobadian, A., O'Regan, N.: e-Leadership through strategic alignment: an empirical study of small- and medium-sized enterprises in the digital age. J. Inf. Technol. **31**(2), 185–206 (2016)
37. Agrawal, A., Catalini, C., Goldfarb, A., Luo, H.: Slack time and innovation. Organ. Sci. **29**(6), 1056–1073 (2018)
38. Grass, A., Backmann, J., Hoegl, M.: From empowerment dynamics to team adaptability: exploring and conceptualizing the continuous agile team innovation process. J. Prod. Innov. Manag. **37**(4), 324–351 (2020)

Requirements for a Digital Business Ecosystem Modelling Method: An Interview Study with Experts and Practitioners

Chen Hsi Tsai(✉) , Jelena Zdravkovic , and Janis Stirna

Department of Computer and Systems Sciences, Stockholm University, Stockholm, Sweden
{chenhsi.tsai,jelenaz,js}@dsv.su.se

Abstract. Digital Business Ecosystem (DBE) supports organizations to collaborate and combine their expertise to stay competitive through ICTs. Being beneficial to the organizations, DBEs are also complex and more difficult to manage. Through visual analysis and knowledge explication, modelling can increase the transparency of DBEs and improve the understanding of DBEs through multiple perspectives. However, current scientific literature suggests a lack of methodological guidance for modelling in support of the analysis, design, and management of DBEs. The purpose of this interview study was to elicit requirements and needs for developing a holistic method of modelling DBEs. Semi-structured interviews were conducted to explore and understand DBEs from an industrial and experts' viewpoints with 11 participants from different business domains. The empirical data was analyzed using Thematic Analysis, which lead to 30 requirements, categorized into four classes, namely, requirements for model creation, analysis, management of DBE, as well as cross cutting requirements addressing the overall establishment and use of the modelling method.

Keywords: Digital business ecosystem · Requirements · Enterprise modelling

1 Introduction

Digital Business Ecosystem (DBE), a new type of collaborative network among organizations, and its formation has been accelerated by the Internet with massive amount of online collaborations. In the context of a DBE, organizations and individual actors cooperate for product and service delivery or utilization.

The notion of DBE is based on Moore's concept of Business Ecosystem which emphasizes evolution and co-evolution in a complex system involving organizations and individuals of an economic community [1–3]. As a biological metaphor emphasizing the interdependent actors in the ecosystem, the digital aspect of DBE considers a technical infrastructure distributing any useful digital representations, such as software applications, services, descriptions of skills, laws, etc., while the business aspect is similar to Moore's idea.

© Springer Nature Switzerland AG 2021
R. A. Buchmann et al. (Eds.): BIR 2021, LNBIP 430, pp. 236–252, 2021.
https://doi.org/10.1007/978-3-030-87205-2_16

As compared to traditional multi-actor business models that are manufacturer, retailer, or franchise centered, a DBE possesses some unique characteristics, including heterogeneity, symbiosis, co-evolution, and self-organization, which enables it to incorporate different business domains and diverse interests in its digital environment [2, 4]. Heterogeneity denotes the constitution of DBEs with different features and types of companies, organizations, and actors. Symbiosis emphasizes the relationships among DBE actors that depend on each other in particular ways [5] and get benefits or co-create greater value through the interdependencies. Co-evolution refers to the collective transformation of DBE actors from one stage to another, especially their capabilities and roles, while facing opportunities and threats. Self-organization indicate DBEs' ability to learn from their environments and accordingly respond by adjusting to the changing contexts [2]. Hence, DBE is highly valued as a novel collaborative approach to meet elaborated business requirements by enabling capabilities across the involved members, to meet each of their goals and to leverage the offered and desired resources among actors and industries [5].

Despite this beneficial aspect of a DBE for the involved actors, its complexity, due to many correlated or dependent interactions and interrelationships among actors, often makes it difficult to manage [5]. Conceptual modeling and Enterprise Modeling have proven to be useful in dealing with complex problems in organizational settings [6]. Consequently, modelling can be seen as a good way to deal with the complexity of DBEs, because it allows capturing and documenting the important parts of it and to enhanced the level of abstraction for analysis and decision making. There are, however, challenges. The current scientific research suggested that there is a lack of means of model development, analysis, measurement, and management supporting DBEs [2, 7]. Furthermore, an ongoing systematic literature review, conducted by the authors, reveals a number of issues concerning the currently existing modelling approaches for analysis and design of DBEs. For example, the current approaches do not cover all necessary DBE concepts in their modelling language constructs. In relation to this, they do not support the capturing of the multiple perspectives of a DBE, which leads to difficulty in understand the complexity of DBEs in a holistic way. Moreover, many of these approaches include essential and clearly identifiable perspectives for initiating a DBE, such as the actor perspective, while other perspectives, equally important, e.g., the policy perspective, are currently under-addressed. Additionally, a scarcity of tools for supporting DBE modelling methods is also observed.

To this end, we consider that there is a need of a holistic (integrated and multi-perspective) modelling method which provides explicit guidance on the analysis, design, and management of DBEs. Therefore, the goal of this paper is to present an interview study devoted to eliciting industrial requirements for a DBE modelling method. This study is a part of an ongoing design science research project.

The rest of this paper is organized as follows. Section 2 gives background on modelling methods and the current research in DBE modelling. Section 3 presents the methods of conducting this semi-structured interview study. In Sect. 4, results of the thematic analysis are reported together with the elicited requirements. Section 5 provides a discussion on the findings and conclusion remarks.

2 Background

According to Karagiannis and Kühn [8], a modelling method consists of three parts: 1) a modelling language, including syntax, semantics, notation, and language construct (metamodel); 2) a modelling procedure, describing steps and guidelines for applying the language to create valid models; and 3) modelling algorithms, executed on the modelling language. Additionally, we consider the tools for supporting modelling methods important and closely related to modelling methods.

Using a modelling method as an approach, which aims to simplify complex systems, can be an appropriate way for describing DBEs and addressing their complexity. Nevertheless, the current state of the art suggests that the area of DBE modelling methods lacks practicable solutions that address the diverse aspects of a DBE as well as support the establishment and management of the DBE throughout its lifecycle. We have investigated this topic in an ongoing systematic literature review, where 3509 studies were retrieved and 63 included in analysis. The review has, so far, revealed that few scientific studies proposed a comprehensive conceptual modelling method for DBEs. Not many studies proposed a method consisting of modelling languages and guidelines or procedures for modelling DBEs. A few examples of studies which suggested a more comprehensive method were the Methodology of Business Ecosystem Network Analysis (MOBENA) in [9], the methodology for modelling interdependencies between partners in a DBE in [5], and the approach for modelling and analyzing DBEs from the value perspective in [10].

Also, many of these studies did not include all essential elements of DBEs in their modelling language constructs. *Actor, role,* and *digital component* were the elements commonly being addressed, whereas *capability,* as an example, was neglected. For instance, Aldea's [10] study was one of the studies which successfully included most of these essential elements (*actor, role, capability, relationship,* and *digital component*). The number of studies suggesting a modelling procedure for the proposed method was also low. A novel contribution was seen in [11] where a top-down policy-based DBE modelling approach with its procedure was proposed. Among the proposed procedures, the most prominent steps of these procedures, such as *identifying actors, roles, or digital components,* were in accordance with the commonly included modelling elements.

One study, as we have observed so far, envisioned a modelling tool (OmiLAB) aimed to support DBE design and management [7]. This indicates the insufficiency of tools for supporting DBE modelling methods.

3 Methods

This study is part of a design science research (DSR) project which aims to develop a management framework for the resilience of a digital business ecosystem. According to the DSR guidelines and research process [12, 13], this exploratory semi-structured interview study contributes to the iterative steps of problem identification and requirements analysis for the design artefact – a modelling method for DBEs. The exploratory semi-structured interview followed an interview guideline with a set of predetermined open-ended questions. The questions were structured into four blocks, namely general and background, planning and designing of DBE, deployment and operation, and monitoring. Each interviewee received one 1–1.5 h-long interview session (except interviewees P1 and P2 as shown in Table 1) via digital meeting (Zoom, or similar). In the beginning of each session, a brief summary of the concept of DBE was exchanged with the interviewee. Based on the informed consent obtained prior the interview process, the sessions were recorded for further analysis purposes.

Purposive sampling was conducted in order to recruit study participants who are practitioners or experts. The reason was that practitioners have been working with or taking part in DBEs and thus could possess industrial knowledge and experiences which are related to DBEs and similar multi-actor constellations/networks, whereas the experts could have expertise in specific fields which are considered important in DBE studies. Table 1, 11 interviewees participated in this study.

Table 1. Description and identification codes (id.) of the interviewees

ID	Description of interviewee
P1	CEO of a Swedish digital platform company as part of a DBE in healthcare
P2	CEO of a global and diverse network of professional women and experts, dedicated to creating a platform for economic empowerment
P3	Project Manager of a leading Research and Technological Development Centre in Europe
P4	Principal and Head of Information Security and Governance Services of a cybersecurity consultancy firm
P5	Enterprise Architect of a European high-performance full-service provider for municipal utilities, energy traders and other utility companies
P6	Head of IT Excellence of a European telecommunications operator
P7	Lead Enterprise Architect of a Swedish multinational telecommunications company and mobile network operator
P8	Senior Business Technology Analyst of a leading company of IT Management in the Nordics and pioneer of Business Technology management in Northern Europe
P9	Senior project manager of a European telecommunications operator
P10	Analytical Lead of an American multinational technology company that specializes in Internet-related services and products
P11	CEO of a British multinational company with a vision to develop dynamic software systems designed to adapt to the changing challenges businesses face

For data analysis, a thematic analysis approach [14] was adopted to annotate and collate the interview transcriptions and summarize the findings using the qualitative data analysis software NVivo. The codes and categories were based on the empirical data (interview transcriptions). In order to abstract to a higher level, the thematizing of the categories was conducted with the support of theoretical proposals on the resilience perspective of DBE [15, 16] and the existing DBE frameworks [5, 17–19].

4   Results

The themes, categories, and codes emerging through the thematic analysis of the qualitative data are presented in this section. From seven categories and their corresponding
25 codes, two themes, namely DBE Fundamentals and DBE Resilience, transpired.
Together with the analysis, requirements of a DBE modelling method informed by the
codes are suggested. In total, 30 requirements were elicited. The establishment of these
requirements was facilitated by the NVivo tool as shown in Fig. 1.

In Sect. 4.1, the theme DBE Fundamentals is elaborated with its four categories -
boundaries and scope; *communication*; *collaborative digital environment*; and *adherence to rules*. In Sect. 4.2, the theme DBE Resilience is elaborated with its three categories
- *adjustment to changes*; *agreement and alignment*; and *management*.

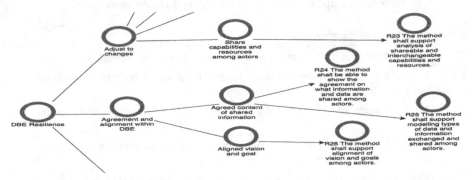

Fig. 1. Establishment of the requirements supported by codes in NVivo (part of the requirements
as an example)

4.1   DBE Fundamentals

Boundaries and Scope. The boundaries and scope of a DBE can be unclear as the
interconnections among different actors and DBEs can be infinite in the digital realm,
resulting in a huge DBE which includes many other DBEs. To scale down a DBE and
delimit the boundaries and scope around it is therefore crucial. The boundaries concern
which actors are in a DBE, whereas the scope pertains to business activities and values
of the DBE with respect to industrial sector and business focus.

Table 2 shows the two codes of the category, which requirements they motivate, as
well as the selected citations from the interviews for the codes.

Understanding the specific focus of business on which actors in a DBE are collaborating and conducting can support defining the scope of the DBE (as no. 1 in Table 2). A
way to carry out the process of delimiting a DBE is through identification of significant
actors or organizations and the roles they are playing within the DBE. It is also important
to identify other actors who participate in the value exchange and are considered part
of the DBE (as no. 2 in Table 2). Additionally, defining the responsibilities according to
involving actors and their roles is critical (as no. 3 in Table 2).

Table 2. Codes and examples of citations of the category *boundaries and scope*

Code	Citation
Boundaries defining through actors and responsibilities (motivates R1. And R2.)	no. 1 *"In the ecosystem, you need to have people that do business together... Not any kind of business" (P11)*
	no. 2 *"So you have to find the imaginator or the driver...And you have to define the benefit carriers or the people who draw value from the DBE because they're also part of that DBE... you use the imaginator, or the driver...as the starting point to carve out the beginning and the end, sort of the boundaries of the DBE" (P4)*
	no. 3 *"Another interesting thing with them was that they had some they had defined roles and responsibilities based on these networks (DBEs)...that was a very important part of it." (P8)*
Without geographical restrictions (motivates R3. and R4.)	no. 4 *"There are no (geographical) boundaries...It's virtual. It's in the digital world...That geographical digital realm, which is without (geographical) boundaries." (P6)*

Another nature of a DBE is that it is not restrained geographically. This means that an actor in a DBE can be anywhere geographically (as no. 4 in Table 2). Based on the codes collated in this category, four requirements (R1.–R4.) are elicited as follow:

- R1. The method shall aid the process of delimiting scope and boundaries of a DBE.
- R2. The method shall support assignment of DBE roles and responsibilities to actors.
- R3. The method shall support inclusion of actors regardless of geographical locations.
- R4. The method shall support exposure of a DBE regardless of its geographical location.

Communication. In the interorganizational settings of DBEs, communication, meaning the exchange of information between organizations or actors, is considered crucial.

The two codes belonging to this category are shown together with selected citations in Table 3.

Communication within a DBE and the channels, such as the platform and line used, are essential (no. 1, no. 2 and no. 3 in Table 3). The electronic way of message exchange supporting by machine interpretable semantics is needed when communicating mass amount of organizational information (no. 4 in Table 3). Relating to these semantics, standardization as the establishment and implementation of standards for communication is needed to build up a DBE. Endeavors to create such standards which can be used for DBEs have been on the governmental and scientific agenda (no. 5 in Table 3). For this category, two requirements (R5. and R6.) are elicited as follow:

Table 3. Codes and examples of citations of the category *communication*

Code	Citation
Communication channels (motivates R5.)	no. 1 *"Communication is the blood in DBEs."* *(P4)*
	no. 2 *"...in the definition of DBEs there was their communication platform...this communication platform was very important."* *(P8)*
	no. 3 *"...the line of communication, the type of information, the type of data that people will be exchanging will come into picture in this metamodel."* *(P11)*
Common way of communication (motivates R6.)	no. 4 *"When we speak about communication between organizations...because of the mass of information, the mass of messages...must be electronically...there must be formal semantics. Messages must be based on formal semantics in which they can be interpreted by a machine and not..."* *(P5)*
	no. 5 *"To be able to build such an ecosystem is standards, you know, you need standards...There are a number of European projects, they're trying to build a standard for exchanging data and information."* *(P11)*

- *R5. The method shall support modelling of communication channels among actors.*
- *R6. The method shall support modelling of commonly agreed form of communication among actors.*

Collaborative Digital Environment. A mixture of actors, processes, information, products, services, and infrastructures exist in the digital environment of a DBE. Table 4 shows the codes related to this category and some selected citations.

The digital environment is considered distributed and decentralized in which the collaborating actors remain autonomous to a certain extent (no. 1 and no. 2 in Table 4). Within the environment, the dual nature of relationships among the varieties of actors is recognized. Actors who are competitors can also be collaborators (no. 3 in Table 4).

Being inclusive and having low entry barriers, the environment supports actors to join, maintain, and collaborate with each other for value co-creation (no. 4 in Table 4). The integration of processes among actors within the digital environment is challenging but important (no. 5 and no. 6 in Table 4).

A healthy collaboration within the environment should be ensured. A way to address this is to share relevant information concerning the collaborations with actors in the environment (no. 7 in Table 4). Value proposition for all actors should also be considered to secure the collaborative property of this environment (no. 8 in Table 4).

Table 4. Codes and example citations of the category *collaborative digital environment*

Code	Citation
Decentralized and distributed (motivates R7.)	no. 1 *"If you bring that paradigm of ecosystem in business environments, then it usually will be distributed." (P11)*
	no. 2 *"I would say it's a very, very decentralized model." (P6)*
Dual nature of relationships (motivates R8.)	no. 3 *"You have this dual nature, you have two actors that are competing, would like to eat each other... But at the same time, they need to cooperate." (P6)*
Inclusive and low entry barriers (motivates R9.)	no. 4 *"...why I think that those SMEs, as I mentioned, as value creators are also important because when this ecosystem is also inclusive, and have not so hard barriers for entering also for SMEs entrepreneurs..." (P2)*
Integration of actors and their processes through digitalization (motivates R10. and R11.)	no. 5 *"...how we can build and integrate business processes together without forcing the other party to use one of your systems. So, it's all about integration and information exchange and collaboration, but in a digital environment." (P11)*
	no. 6 *"It's a mix of people, processes and infrastructures that come together." (P10)*
Healthy collaboration through mutual understanding (motivates R12.)	no. 7 *"It's really important to ensure the collaboration is healthy...But if the customer sees, you're increasing the price...he will probably decide to change provider. But if he knows that there is this regulation issues...And even if the price is higher... he will think, ok, you're working, I'm not going to change because you're raising the price because it's paying to the EU." (P3)*
Value co-creation and proposition (motivates R13. and R14.)	no. 8 *"Everyone must be winner, otherwise, you're not part of an ecosystem. And here we have the users; we have the payers; we have the financer; we have different solutions and products, we reselling that; we're giving you market, etc., ...it's creating winning value proposition for all parties in the ecosystem." (P1)*

(continued)

Table 4. (*continued*)

Code	Citation
Digital infrastructures shared among actors for innovation (motivates R15. and R16.)	no. 9 *"So it's just taking our customer, usual customer and offering additional service, then you can join forces with another actor...for example, I'm having, as a customer, my web page...I see that there's a new different section...there are now smart home or health benefits, how to reach your health providers via my telco. So yeah, that's not using something new. You join forces and you're using one's company solutions." (P8)*

Multiple digital infrastructures, such as cloud services and web solutions, are jointly used by the actors. The sharing of these digital infrastructures can lead to innovations in new services and products as illustrated by the case where a web-based solution of a telecommunications company was shared with healthcare and smart-home companies (see citation no. 9 in Table 4).

In total, 10 requirements (R7.–R16.) are supported by the codes in this category:

- *R7. The method shall support analysis of an actor's relevant part of a DBE.*
- *R8. The method shall support analysis of relationships and types of relationships among actors.*
- *R9. The method shall support the means and requirements for the accession to a DBE.*
- *R10. The method shall support modelling of actors' processes, capabilities, and resources on a DBE level.*
- *R11. The method shall be able to integrate actors' processes and streamline the overall process on a DBE level.*
- *R12. The method shall support exposure of collaborating processes, capabilities, and resources among actors.*
- *R13. The method shall support modelling of values within a DBE.*
- *R14. The method shall support analysis of existing and future co-created values among actors.*
- *R15. The method shall support modelling of digital infrastructures within a DBE.*
- *R16. The method shall support analysis of existing and future innovation created by shared infrastructures among actors.*

Adherence to Rules. This category concerns DBE actors' adherence to policies and regulations. The codes of this category are shown in Table 5.

Various policies and regulations are enforced on actors of DBEs within different domains (no. 1 in Table 5), especially those industries which are highly regulated by the states, such as telecommunications and utilities in the European Union (EU). There are also general policies and regulations which aim to be applied to all domains, such as the General Data Protection Regulation (GDPR) (no. 2 in Table 5) and a number

Table 5. Codes and example citations of the category *adherence to policy and regulations*

Code	Citation
Policy and regulations (motivates R17.)	no. 1 *"In Europe, the utility domain is strongly regulated...Regulations are focused on the communication between the different organizations in that network." (P5)*
	no. 2 *"GDPR it's a kind of a policy, but it's across domains." (P11)*
Less regulated actors (motivates R18.)	no. 3 *"I think we have the EU data protection regulation, which is really ambitious...DBEs are global, you know, they're they don't see any, any geographical boundaries. Laws and regulations are usually geographically bounded, they have their own specific jurisdiction, right...this creates a little bit of a trouble." (P4)*

of others, e.g., the PSD2 and AML directives governing the banking sector in the EU. However, the nature of DBEs without having geographical boundaries could cause issues in enforcing such policies and regulations on collaborating actors which are outside of the jurisdictional or geographical limits (no. 3 in Table 5). The codes collated in this category support two requirements (R17. and R18.):

- *R17. The method shall support modelling of policies and regulations within a DBE.*
- *R18. The method shall support analysis of coverage of policies and regulations.*

4.2 DBE Resilience

Adjustment to Changes. This category concerns the changing nature of a DBE and the means of adjustments. Table 6 shows the six codes of the category and the corresponding selected citations.

A DBE is a dynamic business model constantly changing taking on its own lifecycle once formed (no. 1 and no. 2 in Table 6). This urges the collaborating actors to adopt a fast pace of reaction and decision-making (no. 4 in Table 6). However, this changing nature can be a challenge of understanding and capturing the socio-technical perspectives of a DBE for business purposes (no. 3 in Table 6). It is crucial to keep growing and expanding the framework of understanding and learning of DBEs (no. 5 in Table 6).

To "listen" becomes crucial in such a dynamic collaborative environment, meaning that the actors should be able to gather information about the surrounding contexts, such as trending developments in technology or new decisions made by other collaborating actors (no. 6 in Table 6).

Table 6. Codes and examples of citations of the category *adjustment to changes*

Code	Citation
Evolving and changing contexts (motivates R19.)	no. 1 *"One of the components must deal with somehow this constantly changing nature of DBEs, and that they being alive and out of control of a single person." (P4)*
	no. 2 *"The ecosystems are changing during the changes." (P9)*
	no. 3 *"A new challenge with DBEs is also because, they're changing - the fast-growing innovation and the links they're creating." (P9)*
"Fast" pace – decision-making and reaction (motivates R19.)	no. 4 *"...they had to be really quick in finding out new market needs, and responds quickly to that...something changed in the market, and that was affecting one of these actors, they could decide about it." (P8)*
Expandable DBE framework (motivates R20.)	no. 5 *"So to create this framework or framework of learning...I think it's vital to also get this growing..." (P1)*
"Listen" as a way to adjust (motivates R21.)	no. 6 *"To be able to listen, and to be able to do assessment and analysis all the time, on what is going on in the world. To see the technical development, to understand how your clients and other actors in this DBE reacts to certain decisions you take." (P4)*
Join forces and involve new actors (motivates R22.)	no. 7 *"My understanding about the DBE here is that there is diversity." (P9)*
	no. 8 *"There are lots of changes going on. And the ecosystem is also changing. As they respond to changes, you involve new actors simply in the system, or you make new alliances or you connect to some other." (P6)*
Share capabilities and resources among actors (motivates R23.)	no. 9 *"I have cells here and they are in terms of Telecom. I have that many cells coverage and speed. If your cell is out...maybe you can hire cells from me and use my cell so that you can supply services to your customer continuously and indefinitely. I mean, there would be probably the good possibility to create kinds of markets, right, and to have resources and services to share - so that your ecosystem is working properly." (P6)*

Diversity in a DBE, in terms of involving and joining forces with new actors, can be a way to adjust to changes but contributes also to the changing of the DBE (no. 7 and no. 8 in Table 6). To share resources and capabilities can be another way for collaborating actors to react to changes and ensure the proper functioning of a DBE. An example is illustrated in the citation no. 9 in Table 6. Five requirements (R19.–R23.) are supported by the codes in this category:

- *R19. The method shall be agile in order to support the dynamics of a DBE and fast decision-making withing the DBE.*
- *R20. The method shall support an expandable knowledge base of the DBE framework and roles.*
- *R21. The method should support identification of possible changes by obtaining relevant data.*
- *R22. The method should support analyzing of future state of a DBE when involving new actors.*
- *R23. The method shall support analysis of shareable and interchangeable capabilities and resources.*

Agreement and Alignment. This category concerns the cohesion of a DBE, meaning the agreement and alignment among collaborating actors. The two codes collated in the category and the corresponding selected citations are shown in Table 7.

Due to the variety of collaborating actors and the dual nature of relationships among them within a DBE, it is important for actors to agree upon the shared information content and each actor's accessibility to it (no. 1 and no. 2 in Table 7). To reach an agreement on this is, nevertheless, a great challenge as most collaborating actors might not be willing to exchange data (no. 3 in Table 7).

The collaborating actors in a DBE are doing a business together and not any kind of business (no. 4 in Table 7), meaning that they should have a common business vision. To align the goals among actors can address the challenge of establishing the collaborative environment of a DBE (no. 5 and no. 6 in Table 7). Three requirements (R24.–R26.) are supported by the codes in this category:

- *R24. The method shall be able to show the agreement on what information and data are shared among actors.*
- *R25. The method shall support modelling types of information exchanged and shared among actors.*
- *R26. The method shall support alignment of vision and goals among actors.*

Management of DBE. This category is about the monitoring and measurement of a DBE for the various collaborating actors in the DBE. There are four codes which are related to the category. Table 8 shows the codes and the selected citations of them.

Depending on collaborating actors' interests, they might want to manage and monitor different aspects or domain specific focuses in a DBE, illustrated in an example of governmental agencies' societal interests (no. 1 in Table 8). However, monitoring of business processes and related information is crucial and should be a concern for all

Table 7. Codes and example citations of the category *agreement and alignment*

Code	Citation
Agreed content of shared information (motivates R24. and R25.)	no. 1 *"...some information could be shared and you want to share, it's good to share. But because of this competitive nature of relationship, you don't want to share everything...we would need to know what we can share and know what we don't want to share." (P6)*
	no. 2 *"...you can agree with, with a partner that you only want to share...a silly thing, I mean, their name or surname, but you don't want to interchange the rest of information, but only with one company but with the rest, no." (P3)*
	no. 3 *"This is the biggest problem in this from my experience, people are not willing to share and exchange data." (P11)*
Aligned vision and goal (motivates R26.)	no. 4 *"In the ecosystem, you need to have people that do business together... Not any kind of business" (P11)*
	no. 5 *"...to be honest, well, there is always a challenge, but when you find the common goal and common interest, then the challenge is much, much lower..." (P6)*
	no. 6 *"...we need to set the goal for all the actors so we are on one page and know that our goal is to support a company or an actor..." (P9)*

actors in a DBE. A potential challenge lies in how to monitor the overarching processes at a higher level – the DBE level (no. 2 and no. 3 in Table 8).

Concerning the digital environment and the infrastructures jointly used by collaborating actors for delivering services, KPIs and indicators should be monitored on each of these connection points between the actors and platforms (no. 4 in Table 8). When new actors are joining the environment of a DBE, it is important to assess if they are qualified, in terms of their quality and other properties in accordance to the DBE (no. 5 in Table 8). The codes of the category support four requirements (R27.–R30.):

- *R27. The method shall support management based on actors' domain specific focus areas and interests.*
- *R28. The method shall support management and monitoring of various levels of processes and monitoring of the overall process at a DBE level.*
- *R29. The method shall support monitoring of actors' KPIs and indicators.*
- *R30. The method shall support analysis of various data concerning actors' properties and performance.*

Table 8. Codes and examples of citations of the category *management of DBE*

Code	Citation
Different interests in managing DBE (motivates R27.)	no. 1 *"Maybe (national) telecom agency, for example...their primary interest is social interest...that Swedish society has operating network...that people in Sweden have access...others like GDPR agency...They are just concerned about privacy for customers...that customers are protected." (P6)*
Monitoring of all levels of processes (motivates R28.)	no. 2 *"it's structured as a as a level of process with all the information linked to it." (P3)*
	no. 3 *"If you have multiple organizations, I think it's not a problem. And there's a must for every organization to monitor inside the organization...but if to monitor the overarching process, it's much more difficult because every company would like to keep their information privately. But it's a must to...monitor inside your organization...that it's a must." (P5)*
KPIs and indicators (motivates R29.)	no. 4 *"Monitoring of indicators or some KPIs should be done on every relationship here – the connecting points...that has something to do with the digital platforms..." (P7)*
Qualification of actors (motivates R30.)	no. 5 *"Qualification of new actors when they are joining an ecosystem is needed...it should be done..." (P1)*

5   Discussion and Conclusions

In this study we have applied a qualitative approach with semi-structured interviews and thematic analysis to elicit the requirements for a method for modelling DBEs. A list of 30 requirements has been elicited based on the codes generated during the thematic analysis. Four different classes of requirements have been analytically observed and derived from this list, namely *crosscutting, model creation, analysis,* and *management* related. These four classes together form a framework for the aspects which the modelling method as a design artefact should support when studying a DBE.

Table 9 shows how each requirement from the thematic analysis (TA) categories (c.f. Sect. 4) fits into the four requirement classes. We observed that two TA categories have requirements fitting in the *crosscutting* requirement class. For the class *management*, requirements exist in only three TA categories. The four classes of requirements are further described in the following paragraphs.

The class of *crosscutting* requirements concerns the overall establishment and process of the modelling method. This means that all other requirements should support, brace, and not contradict with each of these cross-cutting ones. The requirements

Table 9. Cross-matching of the 30 requirements in their belonging thematic analysis (TA) categories and requirement classes

Requirement class/TA category	Crosscutting	Model creation	Analysis	Management
Boundaries and scope	R1, R3	R2		R4
Communication		R5, R6		
Collaborative digital environment		R10, R11, R13, R15	R7, R8, R11, R14, R16	R9, R12
Adherence to rules		R17	R18	
Adjustment to changes	R19, R20		R21, R22, R23	
Agreement and alignment		R24, R25, R26	R26	
Management of DBE			R30	R27, R28, R29

included in this group are R1, R3, R19, and R20. For example, supporting expandable knowledge base for the DBE framework and roles (R20) as a modelling method concerns the establishment of the method generally and thus should be supported by all other requirements. This class of requirements can support the development process when outlining the general characteristics of the modelling method, such as delimitation of scope and boundaries (R1).

Model creation requirements are about the capturing and documenting of concepts in DBE with support of the modelling method. R2, R5, R6, R10, R11, R13, R15, R17, R24, R25, and R26 are included in this class. Concepts such as DBE actor, role, responsibility, communication, process, resources, capability, value, digital infrastructure, policy, regulations, agreement, vision, goal, and type of information have been elicited in the requirements. This requirement class facilitates the development of the DBE modelling method so that necessary concepts needed for capturing, documenting, and modelling are included in the method. To clarify the concepts of policy and regulations, it is important to note that regulations are distinguishable from policy in a way that they are often made and imposed by public agencies as restrictions, with the effect of law, on activities within a community [20]. Policy, on the other hand, are usually rules that are made by companies, organizations, groups, or governments in order to carry out a plan and achieve certain aims and goals. In the context of a DBE, there could be external policies to which the DBE needs to adhere, but also but also internal policies in the DBE to which the actors need to adhere.

The class of *analysis* requirements concerns the conducting of analytical actions, often improvement of the DBE in study, using the modelling method. This class contains R7, R8, R11, R14, R16, R18, R21, R22, R23, R26, and R30. Many of these requirements will need to be further refined with analytical or measurable elements in the later stages of the design science research. For example, regarding R23, the analysis of capabilities and resources could be about their performance, sufficiency, and properties determining the interchangeability. Regarding R30, the analysis of actors' performance and properties could be the aspects of effectiveness, efficiency, or domain expertise. These elements shall be concretized and incorporated later with validation of the initial prototype of the modelling method as the design artefact. For R11 and R26, we argue that it belongs to both the documentary and analytical types since the documenting of process, vision, and goal in models enables the analysis of process integration and vision/goal alignment. Vision differs from goals in that it is a collection of goals on a strategic level. Based on this difference, alignment of both goal and vision are equally important as a scenario in a DBE can be that the overall visions among actors are aligned, whereas, on a more detailed level, the goals are conflicting. The *analysis* class of requirements supports developers in identifying the essential analytical use of the models as part of the DBE modelling method during the development process of it.

The *management* class of requirements is about measuring, monitoring, and exposing certain elements, as parts of the management of DBEs during runtime. The class consists of R4, R9, R12, R27, R28, and R29. For instance, being able to, regardless of geographical locations (R4), manage the exposure of certain elements within a DBE (R12) or of the whole DBE should be supported by the modelling method. Management of a DBE means the accessions of actors, monitoring of processes, and measuring of KPIs, bases on specific interests of a domain. This group of requirements assists in developing the DBE modelling method to support DBE runtime management.

A limitation of the study is that the elicited list of the 30 requirements might not be a complete list. Regardless of the possibility of being incomplete, the elicited requirements are considered valuable and valid since they are derived from the interviews with experts and practitioners. It is common that, in a DSR project, the elicitation of requirements is done iteratively. Additional requirements or refined requirements, as mentioned before, will emerge in the subsequent phases of the project when the initial version of the design artefact will be developed and validated.

References

1. Moore, J.F.: Predators and prey: a new ecology of competition. Harv. Bus. Rev. **71**, 75–86 (1993)
2. Senyo, P.K., Liu, K., Effah, J.: Digital business ecosystem: literature review and a framework for future research. Int. J. Inform. Manage. **47**, 52–64 (2019)
3. Senyo, P.K., Liu, K., Effah, J.: A framework for assessing the social impact of interdependencies in digital business ecosystems. In: Liu, K., Nakata, K., Li, W., Baranauskas, C. (eds.) ICISO 2018. IAICT, vol. 527, pp. 125–135. Springer, Cham (2018). https://doi.org/10.1007/978-3-319-94541-5_13
4. Nachira, F., Dini, P., Nicolai, A.: A network of digital business ecosystems for Europe: roots, processes and perspectives. In: Nachira, F., Nicolai, A., Dini, P., Le Louarn, M., Rivera Leon, L. (eds.) Digital Business Ecosystems, pp. 1–20. European Commission, Bruxelles (2007)
5. Senyo, P.K., Liu, K., Effah, J.: Towards a methodology for modelling interdependencies between partners in digital business ecosystems. In: IEEE International Conference on Logistics, Informatics and Service Sciences (LISS 2017), pp. 1165–1170. IEEE (2017)

6. Persson, A., Stirna, J.: Why enterprise modelling? an explorative study into current practice. In: Dittrich, K.R., Geppert, A., Norrie, M.C. (eds.) Advanced Information Systems Engineering. LNCS, vol. 2068, pp. 465–468. Springer, Heidelberg (2001). https://doi.org/10.1007/3-540-45341-5_31

7. Pittl, B., Bork, D.: Modeling Digital Enterprise Ecosystems with ArchiMate: A Mobility Provision Case Study. In: 5th International Conference on Serviceology for Services, ICServ 2017, pp. 178–189 (2017)

8. Karagiannis, D., Kühn, H.: Metamodelling platforms. In: Bauknecht, K., Tjoa, A.M., Quirchmayr, G. (eds.) EC-Web 2002. LNCS, vol. 2455, pp. 182–182. Springer, Heidelberg (2002). https://doi.org/10.1007/3-540-45705-4_19

9. Battistella, C., Colucci, K., Nonino, F.: Methodology of business ecosystems network analysis: a field study in telecom italia future centre. In: De Marco, M., Te'eni, D., Albano, V., Za, S. (eds.) Information Systems: Crossroads for Organization, Management, Accounting and Engineering: ItAIS: The Italian Association for Information Systems, pp. 239–249. Physica, Heidelberg (2012)

10. Aldea, A., Kusumaningrum, M.C., Iacob, M.E., Daneva, M.: Modeling and analyzing digital business ecosystems: an approach and evaluation. In: 20th IEEE International Conference on Business Informatics, CBI 2018, pp. 156–163. IEEE (2018)

11. Biermann, J., Corubolo, F., Eggers, A., Waddington, S.: An ontology supporting planning, analysis, and simulation of evolving Digital Ecosystems. In: 8th International Conference on Management of Digital EcoSystems, MEDES 2016, pp. 26–33. Association for Computing Machinery, Inc. (2016)

12. Hevner, A.R., March, S.T., Park, J., Ram, S.: Design science in information systems research. MIS Q. **28**, 75–105 (2004)

13. Offermann, P., Levina, O., Schönherr, M., Bub, U.: Outline of a design science research process. In: Proceedings of the 4th International Conference on Design Science Research in Information Systems and Technology. Association for Computing Machinery, Philadelphia, Pennsylvania (2009)

14. Braun, V., Clarke, V.: Using thematic analysis in psychology. Qual. Res. Psychol. **3**, 77–101 (2006)

15. Graça, P., Camarinha-Matos, L.M.: A proposal of performance indicators for collaborative business ecosystems. In: Afsarmanesh, H., Camarinha, L.M., Lucas Soares, A. (eds.) PRO-VE 2016. IAICT, vol. 480, pp. 253–264. Springer, Cham (2016). https://doi.org/10.1007/978-3-319-45390-3_22

16. Tsai, C.H., Zdravkovic, J., Stirna, J.: Capability Management of digital business ecosystems – a case of resilience modeling in the healthcare domain. In: CAiSE Forum, LNBIP, pp. 126–137 (2020)

17. Hadzic, M., Chang, E.: Application of digital ecosystem design methodology within the health domain. IEEE Trans. Syst. Man Cybern. Part A Syst. Hum. **40**, 779–788 (2010)

18. Ma, Z.: Business ecosystem modeling- the hybrid of system modeling and ecological modeling: an application of the smart grid. Energy Inform. **2**(1), 1–24 (2019). https://doi.org/10.1186/s42162-019-0100-4

19. Senyo, P.K., Liu, K., Effah, J.: Understanding behaviour patterns of multi-agents in digital business ecosystems: an organisational semiotics inspired framework. In: Kantola, J.I., Nazir, S., Barath, T. (eds.) AHFE 2018. AISC, vol. 783, pp. 206–217. Springer, Cham (2019). https://doi.org/10.1007/978-3-319-94709-9_21

20. Selznick, P.: Focusing organisational research on regulation. In: Noll, R.G. (ed.) Regulatory Policy and the Social Sciences, pp. 363–364. University of California Press, Berkeley (1985)

Author Index

Printed in the United States
by Baker & Taylor Publisher Services